TRAVELING THE SEA OF NIGHT:

UNDERSTANDING TIME AND DISTANCE IN THE UNIVERSE

TRAVELING THE SEA OF NIGHT:

UNDERSTANDING TIME AND DISTANCE IN THE UNIVERSE

JOSEPH R. CARUSO

BEAR PUBLISHING
Yreka, California

DEDICATED TO EVERYONE WHO, LIKE ME, EVER LOOKED UP AT THE STARS AND WONDERED HOW FAR AWAY THEY WERE—AND HOW WE COULD POSSIBLY KNOW.

About the Author

Astronomer Joseph R. Caruso was the physical science technician and telescope operator for the Smithsonian Astrophysical Observatory at Harvard for nearly twenty years. He operated its 61-inch telescope (he made nearly half, or almost 64,000, of that institution's Echelle Observations on the device), and he also operated the observatory's 84-inch radio telescope for the Paul Horowitz Planetary Society's SETI (Search for Extraterrestrial Intelligence) program. He has also taught college-level astronomy as well as adult education classes, spoken at numerous "star parties" and other astronomical events, lectured at planetariums, and conducted tours at the Oak Ridge Observatory. The 2001 recipient of the Astronomical Society of the Pacific's Las Cumbres Amateur Outreach Award "for contributions to astronomy education and outreach in the Greater Boston area for more than twenty-five years," Mr. Caruso continues to make numerous community presentations, and he has also published in the field in notable journals/magazines, including *Astronomy*, *Sky & Telescope*, *Selenology Today*, *Deep Sky*, *Publications of the Astronomical Society of the Pacific*, and *The Astronomical Journal*.

Table of Contents

Photos/Diagrams/Tables

Photo 1: Planisphere
Photo 2: Models
Photo 3: B.C. Cartoon
Photo 4: The Moon through a small telescope
Photo 5: The Moon through a large telescope
Photo 6: Stars as Tomatoes
Photo 7: The Pleiades
Photo 8: The Perseus OB3 Association
Photo 9: M4 in Scorpius
Photo 10: The Milky Way superimposed on a map of the United States
Photo 11: The Small Magellanic Cloud
Photo 12: The Great Andromeda Galaxy—M31
Photo 13: The Colliding Spiral Galaxies of Arp 274
Photo 14: Coins and ibuprofen tablets representing the Local Group
Photo 15: The Virgo Cluster
Photo 16: The Hubble Deep Field
Photo 17: Shadow of the Sun
Photo 18: Moon Shadows—Sun Overhead
Photo 19: Moon Shadows—Sun at Angle

Diagram 1: Celestial Sphere—North Pole View
Diagram 2: Celestial Sphere from the United States
Diagram 3: View from the Equator
Diagram 4: Earth as it appears to us and to a giant
Diagram 5: Earth's shadow during lunar eclipse

Photos/Diagrams/Tables

Acknowledgments

First of all, I would like to thank William Scheniman, who listened patiently over the years to my ramblings on the phone and asked many pertinent questions during observatory tours. His comments and questions led greatly to the expansion of the contents of this book. I would also like to thank George Atamian, who gave me my first chance to become involved in astronomy, at Talcott Mountain Science Center, and Dr. Arthur Upgren at Wesleyan University, who was my mentor and taught me a lot about parallaxes and the mathematical basis of what we know.

The book would never have become a reality without the expertise and patient work of my editor, Mike Slizewski. His professionalism was crucial to untangling my sometimes less than lucid sentences and punctuation. Special thanks to Steve McKinley, who translated my crude renditions into attractive and meaningful diagrams.

Lastly, I would like to acknowledge the encouragement, help, and typing skills over the years of my wife, Karen.

Notwithstanding all of the help I have received, any errors and misinterpretations are my own.

PREFACE

The Universe seems incomprehensible. I suspect that even some astronomers, while they talk about light-years and megaparsecs, can't really envision the distances.

And what is distance? When you ask someone how far it is from Paris to Amsterdam (or Boston to New York), they often answer "three hours." But this is a time, not a distance—three hours by car, airplane, walking?

Astronomical distances are far more complicated. What if the cities were moving apart from each other as you were traveling between them?

So at the largest scales, astronomers have to deal with "look back" times or "comoving distances." And even on the smallest scales, when an astronomer says that Alpha Centauri is 4.3 light-years away, this seems comprehensible (it's not a big number), but what they are really saying is this—imagine that you could travel at 1,080,000,000 kilometers per hour (669,000,000 miles per hour) for 4.3 years. That's how far away the closest star is.

To make it even more confusing, most people's idea of interstellar distances have been formed by Hollywood science fiction movies, which not only ignore the relativistic effects of light-speed travel (mass increase, length compression, and time dilation) but make it look like it's just a matter of going faster than we can today—no big deal. The fastest spaceship humanity has sent into interstellar space, the Voyager 1, is going at about 64,100 kph (40,000 mph)—that's 1/16,000th the speed of light. At that speed it would take it 70,000 years to get to Alpha Centauri (the closest star)—interstellar travel *is* a big deal.

Another problem that this book will deal with is "unlearning." Everyone has a perception of what the Universe looks like. It may be completely incorrect, but it seems that it is human nature to have at least some rudimentary picture of where we are. It may be as simple as perceiving the Earth as a flat plate that the Sun, the Moon, the planets, and the stars all revolve around, fixed to a big glass sphere that we are inside of. Don't laugh—it's a powerful idea that took thousands of years to demolish, and don't we still use the term sunrise and sunset to describe the passage of days?

Unlike biology, where some knowledge is necessary to your survival or you might walk out in front of a bus and think you'll survive, one can go through life knowing virtually nothing about astronomy. Unless you are a farmer that needs to know when the seasons will change, the only really important fact you need to know is that sometimes it's light and sometimes it's dark.

But outer space is closer than you may think. An hour's drive straight up would take you there. Legally, space begins at 100 km (62 miles) above us—the so-called Kármán line.

Distance is everything. Without knowing how far away an object is, we don't know anything about its nature. Once not long ago many astronomers thought the great galaxy in Andromeda was just a nascent solar system forming near us, not another Milky Way galaxy 2.4 million light-years away. Once the distance was discovered in the 1920s, its true nature was revealed.

Every chapter will also at least touch on *how* we know what we know. If you travel to other countries, you will find a great diversity of opinion concerning subjects like art, religion, philosophy, or politics, but Japanese, Indian, British, or American astronomers all have pretty much bought into the same scientific ideas. Why is that? Well, science, unlike other areas of human endeavor, is verifiable or falsifiable, so when someone comes up with a radical theory (some of which have proven to be true), it is quickly checked using the same tools available everywhere. That's why a large European telescope looks just like a large Chinese telescope. We all live in the same physical Universe. Science is self-correcting. The interplay between theory and observation has made it the most powerful idea that human beings have ever invented.

Everyone seems to have forgotten Francis Bacon, but his book *Novum Organum* ("New Method") changed the world forever. Whenever the theorists conceive a new theory, the observers rush to either prove or disprove it. Sometimes a theory, while intriguing, can't be proved or disproved because the instruments are just not sophisticated enough yet. And sometimes a new instrument is invented that allows scientists to observe something never seen before, so the theorists have to reevaluate their concepts. I think this is why there are so few real "breakthrough" discoveries made at the large expensive observatories; telescope time is precious, so you can't just apply for time on the 4-meter telescope at Kitt Peak National Observatory by saying, "I'd like to use the telescope because I might make a great discovery." Not that they don't contribute immensely to our database and are vital to confirming or falsifying current theories. Of course they do. It's not the data or facts that are important; these change all the time. At one time, it was a "fact" that we lived in a static and unchanging Universe. New instruments (the 100-inch telescope at Mount Wilson Observatory) and a new theory (relativity) changed the facts.

It's the method that is important. Making models seems to me to be the only way that we can begin to comprehend just how vast in space and time the Universe is. When I did the tours at Oak Ridge Observatory, I showed thousands of children and students all the sophisticated telescopes and instruments that we used, but from the feedback that I received, the Solar System model that I made on the grounds, and the burning of a pencil by focusing the Sun's rays on it, made the most lasting impressions on the visitors.

The final dimension this book will deal with is the problem of time. People often come to me, point to a star like Regulus, and say, "I heard that that star is so far away that it may have blown up and we don't know it yet because the light is still coming to us." Aside from the fact that Regulus is not going to blow up (it's not massive enough or old enough), the light travel time from Regulus, 78 light-years, is such a negligible percentage of its lifetime that it is safe to assume that it is still there. Sort of like when you're talking to someone 10 feet away—yes, you're seeing them as they were some billionths of a second ago, the time it took light to travel from her to you, but it's safe to say that it's worth answering her since the lifetime of human beings is so much longer than the light travel times between them.

Some models we will construct will be actual physical models you can readily make at home from things around the house, and some will be imaginary. I hope both will help you to understand the immensity of the Universe.

One other thing—I am well aware of the incredible complications and exceptions that are necessary when talking about anything in astronomy. Calibrating the different kinds of Cepheid variables or trying to derive the absolute magnitudes of the five different kinds of supernovae have been, and continue to be, a real struggle for astronomers. So in order to move the argument along in many places, in fact in almost all of them, I had to simplify the method and ignore the exceptions. While it's simplified, I hope it's not dumbed down to the point of not being valid. My apologies to real astronomers, but the purpose of this book is to render what is incomprehensible into some sort of meaningful picture.

Along with this you will notice that I do some really radical rounding off of numbers. For example, if a star is $10^{1.6} = 39.8107$ parsecs away, I rounded it off to 40—even the best distances to stars are only accurate to about 5 percent. So if you read somewhere that Spica is 261 light-years away, what you're really reading is that it is around 250 to 270 light-years away, and the further we delve into the

Universe, the more the errors pile up—so you may read that M51, a nearby galaxy, is 19, 23, or 27 million light-years away. This is as good as we can get. Missing by 4 or 5 million light-years sounds like a lot, but that's about as well as we can do right now. It's astronomy, not accounting.

CHAPTER 1

APPARENT VERSUS ACTUAL

One of the first problems to overcome is that what is apparently going on in the sky is not what's actually going on. While not actually wrong, different models of the Universe have more limited usefulness as time goes by. For example, the model that you're standing still, and that all the stars are fixed, moving from east to west, and are attached to a celestial sphere is now known to be incorrect, but it is still a useful model that is used every day by navigators traveling on the Earth. In a like manner, Newtonian physics was a perfectly usable model to calculate the trajectory of the Apollo spacecraft as it traveled to the Moon and back; only when speeds approach the speed of light do we find that the Newtonian model of the Universe has to be discarded and substituted by Einstein's relativistic model.

The star finder, or planisphere, is a very useful model to find your way around the apparent sky. The only information you need to know is the date, the time, and which way north is. In order to find the North Celestial Pole, one only needs to find the Big Dipper and look at the two stars that make up the end of the cup; notice their separation (about 5 degrees), continue that distance five times, and you'll notice a fairly bright star all by itself that marks the North Star.

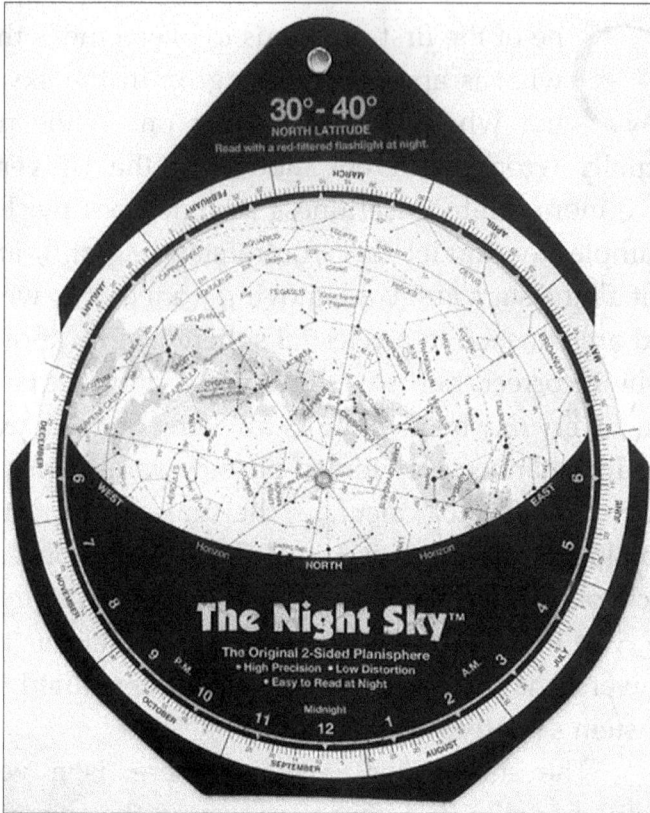

Photo 1: Planisphere (*photo by author*)

Polaris is where the rivet is on the planisphere, so you can't see it. People are sometimes surprised when I point out Polaris (the North Star) to them, thinking that it should be the brightest star in the sky; actually it's the 49th brightest, but it is still easy to find, since it's the only bright star in the area.

Once you have figured out which way north is, turn around and face south holding the planisphere over your head so that the North marker points toward Polaris. Turn the star wheel so that the date and time are correct—for example, October 15 at 10 p.m. Now you are ready to find your way around the sky.

The constellations will then appear as they are on that date and time. All the colorful and sometimes integrated stories that are reflected in the names of the constellations can be ignored for now, as can the fact that the stars that make them up are actually at vastly different distances from us. There are 88 constellations (easy to remember as it is the same number as the number of piano keys or counties in Ohio), but I have found out that if you learn just five of the most obvious ones—one for each season and of course the Big Dipper, which is actually only the most noticeable part of a far larger constellation Ursa Major—you will be well on your way to learning the sky. Let's start in the autumn— say, October 15 at 10 p.m. Set the planisphere for this date and time, face south, and look overhead; you will notice a big square of stars that make up the body of the constellation Pegasus (the Winged

Horse). He's actually upside down, but in space this doesn't matter, and of course for people in the Southern Hemisphere, he's right side up. Now turn the wheel so that it's 11 p.m., then midnight, and so on; you will notice that Pegasus will move across the sky and finally set in the west at about 4:30 a.m. (Planispheres should be made so they only turn counterclockwise since the Earth rotates toward the east and never backs up.)

By this time in the morning the most obvious constellation, and the one with the brightest stars, is Orion (the Hunter), the one to use for the winter skies. Without moving the planisphere, go back and look at 10 p.m. This is when Orion is the highest in the sky, but only in January at 10 p.m. In a like manner, move Orion to set in the west.

Now the most prominent constellation in the sky is Leo (the Lion), which looks like a backward question mark. It is highest in the sky around April 15 at 10 p.m.

Finally, in July at 10 p.m. you will notice that there are three bright stars in different constellations overhead. They are Vega in Lyra (the Lyre), Deneb in Cygnus (the Northern Cross–the Swan), and Altair in Aquila (the Eagle). (So there's no confusion, I always tell people to think of the constellations as countries and the stars as cities. Every city—the stars—is in a country, and the countries—constellations—cover the whole sky.) These three stars make up the Summer Triangle.

At last, if you move the Summer Triangle to the west, you notice that Pegasus reappears in the east. In twenty-four hours a whole day has gone by, but at a given time (say 10 p.m.) a whole year has gone by.

These are constellations that are visible from the Northern Hemisphere on Earth. In the Southern Hemisphere, not only are there different constellations visible, like Crux (the Southern Cross) and Carina (Keel of the Ship *Argo*), but also the ones near the Equator projected out into space will be upside down. Think of it this way: the whole sphere of the sky has 360 degrees. At any one time you can see one-half of it, or 180 degrees. If you were at the North Pole with Polaris directly above your head, you could only see the half of the sky from the Equator north: 90 degrees + 90 degrees = 180 degrees.

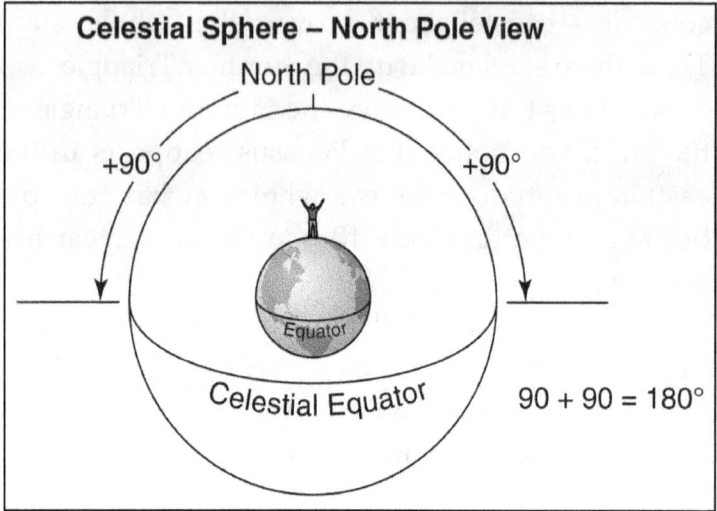

Celestial Sphere – North Pole View

North Pole

+90°   +90°

Equator

Celestial Equator

90 + 90 = 180°

Diagram 1: Celestial Sphere—North Pole View

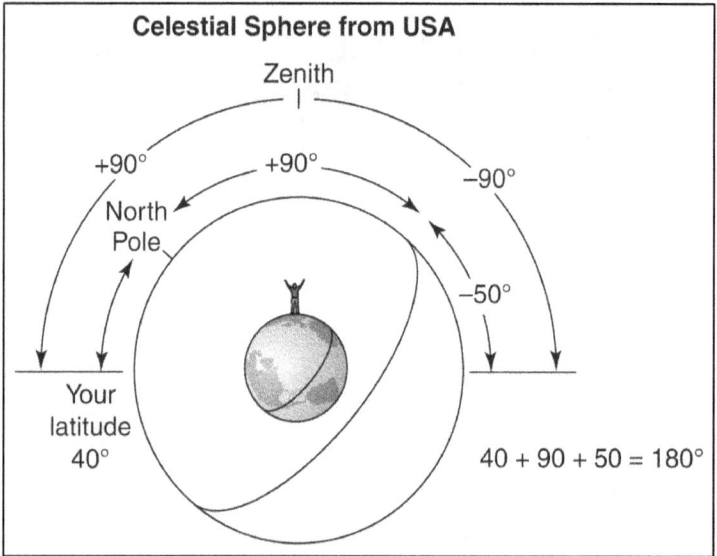

Celestial Sphere from USA

Zenith

+90°   +90°   −90°

North
Pole

−50°

Your
latitude
40°

40 + 90 + 50 = 180°

Diagram 2: Celestial Sphere from the United States

In like manner, if you were at the South Pole, in Antarctica, you would only see half of the sky, the part not visible from the North Pole. Since the United States and Europe are both about halfway between the Equator and the North Pole, we see all of the stars above the Equator and some of the stars between the Equator and the South Pole, depending on your latitude.

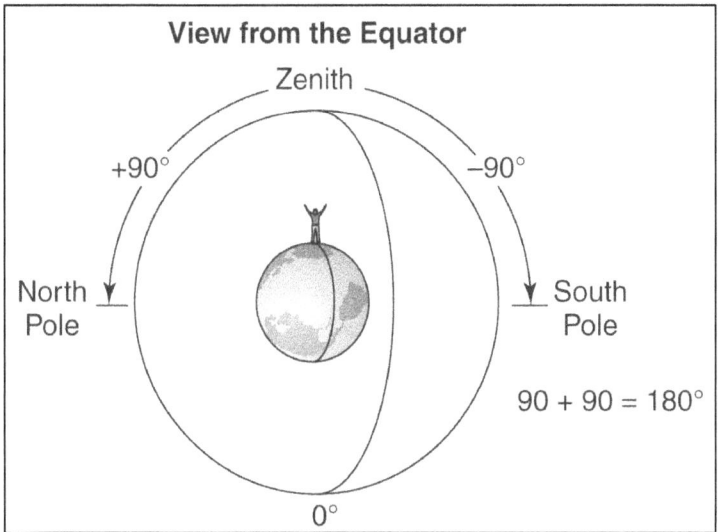

Diagram 3: View from the Equator

Only people living right on the Earth's Equator can see all of the stars at some time of the year.

The point is that how much of the sky you can see depends on where you live on the Earth;

everyone can see half of the celestial sphere (180 degrees) on a given night, but what part of it you can see depends on your latitude, the time of night, whether or not it's cloudy, and if you live where there is so much outdoor lighting that most of the stars are not visible. Since I live in the Midwest of the United States, there is no point in people asking me to see Alpha Centauri, since it's at –62 degrees south of the Earth's Equator projected out into space and the farthest south I can ever see is –52 degrees. People in Florida, the Caribbean, and South America can see it easily.

Whenever I hand someone a planisphere, one of the first questions they ask me is, "Why are the planets and the Sun missing?" The first reason is that, since these move constantly across the celestial sphere, you would have to make a new planisphere every couple of weeks to show where the planets are—in Mercury's case every couple of days. Computer programs like Starry Night do just this, as do apps you can download on your iPhone and Android smartphones and iPad and other tablets. Another way of finding the positions of the planets is to subscribe to the magazines *Astronomy* or *Sky & Telescope*, as they show the sky for the current month. (Actually, the stars are moving too—it's just that they are so far away that new planispheres have to be made only every fifty years or so.)

The second reason is that the Earth is wobbling on its axis like a top spinning down, so the stars apparently shift on the celestial sphere over a long period of time. This is why if you know any constellations at all, you've heard of, say, Leo, Gemini (the Twins), and Taurus (the Bull)—but probably not Camelopardalis (the Giraffe), Lacerta (the Lizard), or Triangulum (the Triangle)—because these are the constellations the Sun apparently is in throughout the year. Astrologers think that this has some effect on your personality and fate, but how balls of gas trillions of miles away can do this is unexplained and ridiculous. Also, due to precession (the wobbling of the Earth's pole), more people are born in the "sign" of Ophiuchus (the Serpent Bearer), where the Sun is from November 27 to December 18, than Scorpius (the Scorpion)—November 20–27—but astrologers conveniently ignore this, along with just about everything else that is going on in the real sky.

The position of the Sun throughout the year can be found on your planisphere by a dashed line labeled "ecliptic," and if you draw a line from the North Celestial Pole to the edge of the wheel where the dates appear (where it crosses the ecliptic), this marks the position of the Sun on that date. For example, on October 15 the Sun (eight and a half minutes away at the speed of light) is in Virgo (the

Maiden)—in fact almost where the bright star Spica is, although it is 240 light-years away in the background and you can't see it, since the Sun is so much brighter that it blocks out all the other stars during the day. Spica is there, but the Sun is so much closer that it only appears brighter—Spica is some 14,000 times more luminous than the Sun. (And all the stars in the sky are up all the time; it's just that some are below the horizon or it's daytime, so you can't see them.)

The modern-day equivalent of the planisphere is a phone app. There are many kinds and levels to accommodate both the beginner and advanced users. Quite a few are free, and the ones that are not are available for a nominal fee. Starmap is one of the easiest to use, because it includes a point-and-identify application. It also allows you to zoom in on planets, clusters, and galaxies. Star Chart is practically the same as above, plus it is free. The Night Sky Lite is also free; it's the one I have on my phone. Like most of the others, it sometimes uses "nonintuitive" symbols; everyone but me seems to know what they stand for. (What does "three gears" mean?) I guess that I missed that day in school. Anyway, it works well enough, if you do not mind being pestered constantly about updating. Distant Suns is probably the most sophisticated of all of them and has just about everything one might want in a smartphone/tablet app, including some really nonuseful items—why would you want to see the back side of Jupiter?

Anyway, there are a lot of other strictly informational apps, and some have fabulous pictures, like the NASA and the Hubble Telescope apps.

While we are on the subject you might want to look into some of the desktop planetariums that are available. I like the aforementioned Starry Night; it came free with a telescope I purchased, and it's great for looking up rising and setting times of the Moon and planets. If you are interested in seeing when the International Space Station will be visible from your location, download heavens-above.com; it has all of the positions of every satellite on any given date and time from your location. The best weather program for astronomers is probably cleardarksky.com, which tells you not only what part of the night will be cloud-free, but also the atmospheric transparency and hours of darkness. For solar events see spaceweather.com—a great site for observing sunspots and anticipating solar activity like coronal mass ejections, and it features polar projections of the Earth, so you can plan when it might be possible to observe auroras (the northern lights)—very educational. Finally, one site that I look at every day is the "Astronomy

Photo Of The Day" (apod.nasa.gov). I first downloaded it in 1995, and I have hardly missed a day since. The pictures cover the whole universe of astronomy—every subject—moons, planets, stars, galaxies, comets, historical pictures, events, space exploration, observatories, and everything else imaginable, submitted by everyone from beginning amateurs to Hubble telescope images.

Before reading any further in this book, you might want to collect the following items: a world globe; a one-sheet map of the continental United States or Europe; two or three tennis balls; some Styrofoam balls of various diameters from 5 to 15 cm (2 to 6 inches); some small balls about 5 mm to 25 mm (a quarter inch to 1 inch) in diameter; map tacks (the round-headed kind); pins; the longest tape measure you can find (30 meters or 100 feet is good); a 152 cm (60-inch) stick; a pencil; various euro coins like the five- and fifty-cent denominations or an American dime and a penny (actually, you'll need 135 billion five-cent euros or American pennies—just kidding, but you'll need a lot); a roll of nonadhesive shelf paper about 15 meters (50 feet) long and 20 to 30 cm (9 to 12 inches) wide; a bottle of ibuprofen tablets (for a couple of reasons); and a planisphere (star finder) or an appropriate star finder app. If you want to do the math, you will also need a scientific calculator that does exponents and logs; that's about the most sophisticated arithmetic that we are going to do.

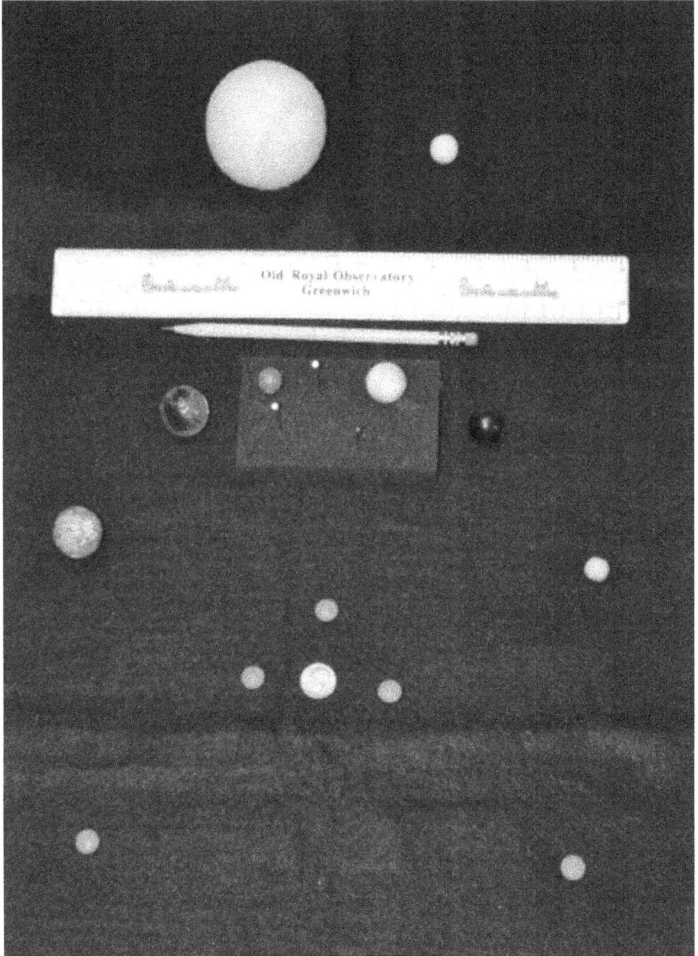

Photo 2: Models (*photo by author*)

CHAPTER 2

THE EARTH AND THE MOON

The first problem in comprehension is trying to imagine the planet on which we live. True, the planet we inhabit is one of the infinitesimally tiniest dust motes in the Universe, but it is still enormous compared to our size, so the first step is to figure out the size and shape of the Earth.

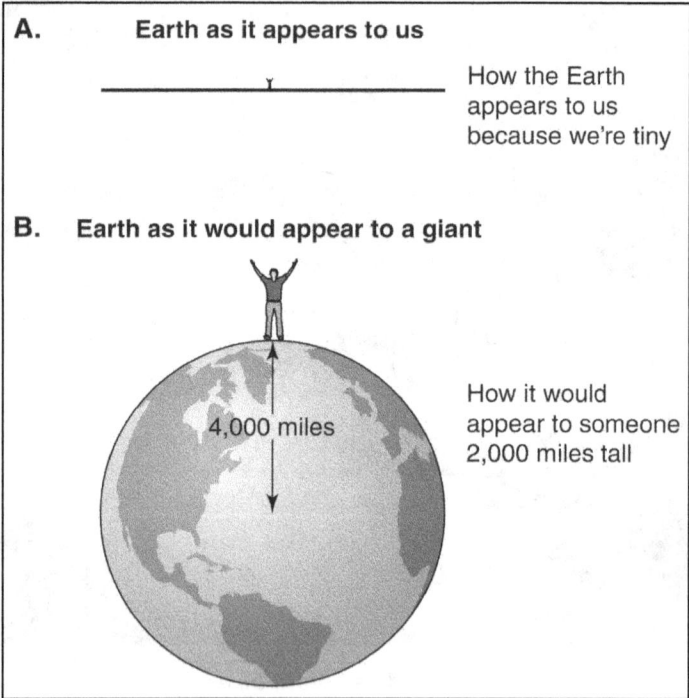

A.   Earth as it appears to us

How the Earth appears to us because we're tiny

B.   Earth as it would appear to a giant

4,000 miles

How it would appear to someone 2,000 miles tall

Diagram 4: Earth as it appears to us and to a giant

Despite the common misconception, educated people have known for thousands of years that the Earth is a sphere; this is easily proved by observing the shape of the Earth's shadow as it passes over the full moon during a lunar eclipse—it always looks like A in Diagram 5, never like B, which it would look like if the Earth were a flat plate. (Also, people living near the sea noticed long ago that the hulls of departing ships disappeared below the horizon before the sails did.) So the Earth is curved, but by how much? Eratosthenes (c. 276–194 BC), a natural philosopher who worked at the library of Alexandria in Egypt, figured it out for us. A little geometry will reveal its size.

Earth's shadow during lunar eclipse

A.

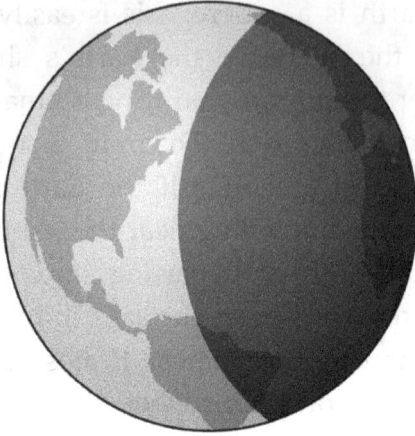

Never like this

B.

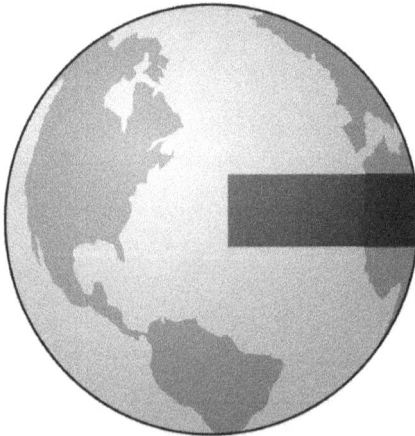

Diagram 5: Earth's shadow during lunar eclipse

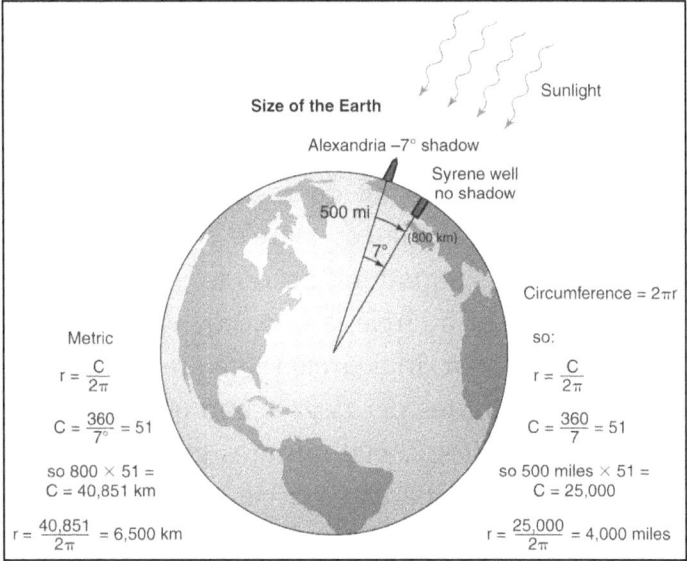

Size of the Earth

Sunlight

Alexandria –7° shadow

Syrene well
no shadow

500 mi

(800 km)

7°

Circumference = $2\pi r$

so:

Metric

$r = \dfrac{C}{2\pi}$

$C = \dfrac{360}{7°} = 51$

so 800 × 51 =
C = 40,851 km

$r = \dfrac{40,851}{2\pi} = 6,500$ km

$r = \dfrac{C}{2\pi}$

$C = \dfrac{360}{7} = 51$

so 500 miles × 51 =
C = 25,000

$r = \dfrac{25,000}{2\pi} = 4,000$ miles

Diagram 6: Size of the Earth

So the next step is the distance to the Moon. People always knew the Moon was the closest object to us except for clouds and maybe meteors, because it moves in front of everything else.

An old "B.C." cartoon is right to the point:

Photo 3: B.C. cartoon (11/12/69) by Johnny Hart
(*By permission John L. Hart FLP and Creators Syndicate, Inc.*)

Johnny Hart's cartoon is amusing but also brings up an important point—if you don't know either the true size or distance to something, it seems impossible to figure out the other. Not true—in ancient Greece, using geometry, the distance to the Moon was calculated. Here's a little geometry refresher: The equation used to calculate the circumference of a circle is $C = 2\pi r$, where r is the radius, so $r = C/2\pi$. In a circle there are 360 degrees, which at 60 minutes per degree and 60 seconds per minute ($360 \times 60 = 21,600$ minutes times 60) equals 1,296,000 seconds of arc in a circle.

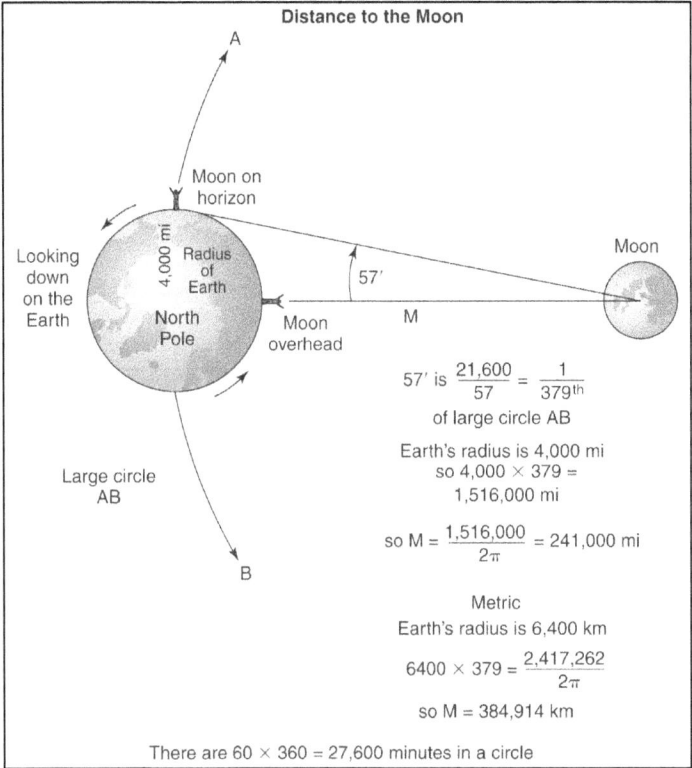

Distance to the Moon

A

Moon on horizon

Looking down on the Earth

4,000 mi

Radius of Earth

North Pole

Moon overhead

57'

M

Moon

Large circle AB

B

57' is $\frac{21,600}{57} = \frac{1}{379\text{th}}$
of large circle AB

Earth's radius is 4,000 mi
so 4,000 × 379 =
1,516,000 mi

so M = $\frac{1,516,000}{2\pi}$ = 241,000 mi

Metric
Earth's radius is 6,400 km

$6400 \times 379 = \frac{2,417,262}{2\pi}$

so M = 384,914 km

There are 60 × 360 = 27,600 minutes in a circle

Diagram 7: Distance to the Moon

Follow the steps in Diagram 7 to calculate distance to the Moon. Let's make a model.

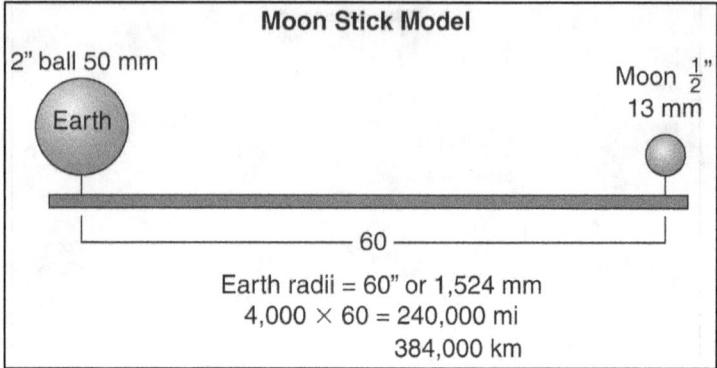

Moon Stick Model

2" ball 50 mm

Earth

Moon $\frac{1}{2}$"
13 mm

60

Earth radii = 60" or 1,524 mm
4,000 × 60 = 240,000 mi
384,000 km

Diagram 8: Moon Stick Model

This simple exercise lies at the heart of what this book is trying to accomplish. Our finite brains can only comprehend such concepts by reducing the scale to human terms.

So it turns out that the Moon is about one-fourth the size of the Earth and some thirty Earth diameters (or sixty Earth radii) away. As you can see, the Moon is a lot farther away than most people think. While you have the Moon stick assembled, go outside someday when it's sunny and you can demonstrate lunar and solar eclipses. Since the Moon is four hundred times smaller than the Sun but four hundred times closer, they appear to be the same size in the sky. Hold the Moon stick so that the shadow of the two-inch ball (the Earth) covers the half-inch ball (the Moon); this is a lunar eclipse and, as you see, it's really easy to do, because the Earth's shadow is so big. But if you try to make a solar eclipse, it's a little more difficult—the Moon's small shadow is hard to line up, because it covers only a part of the Earth. That's why lunar eclipses are readily observed by everyone on the night side of the Earth, but solar eclipses have to fulfill five conditions in order to been seen:

1. You have to be just where the small shadow of the Moon touches the Earth, and since oceans cover 70 percent of the Earth, a lot of eclipses happen out at sea.

2. It has to be when the Sun, the Moon, and the Earth are lined up (this is called syzygy, which would be a great Scrabble word if there were three Ys—actually you can do it with the Y and the two blanks).

3. This is called New Moon, and simultaneously, the Moon's orbit has to be intersecting the ecliptic; this is called "at the nodes."

4. At the same time, the Moon has to be at perigee or near it, which means it is closest to the Earth.

5. Finally, it has to be a clear day.

As I'm sure you're aware, despite our model, the Earth and the Moon aren't really attached to a long stick, so the Moon's orbit is really somewhat eccentric, meaning that it and the Earth orbit around a common barycenter (center of gravity) that is some 1,600 km (1,000 miles) below the surface of the Earth. The Moon goes around the Earth in an egg-shaped path, so sometimes the Moon is closer to us than at other times.

Looking at the model, people often ask me this question: "If we had to get up to 40,000 kph (25,000 mph) to leave the Earth, why did it take three days to get to the Moon?" After all, 384,500 km (240,000 miles) divided by 40,000 kph (25,000 mph) equals about ten hours. Well, the Apollo spacecraft didn't go 40,000 kph all the way to the Moon. It started out that way, but after it had accelerated to 40,000 kph, engines were shut off (most actually thrown away), and the inexorable force of the Earth's gravity started slowing it down again for most of the rest of the way to the Moon, so the average speed to the Moon was only about 3,000 kph. I was always much more amazed at the technology that we developed that managed to slow the Apollo capsule down from 40,000 kph to zero as the spacecraft returned to Earth. I don't know how many times I've heard the media say that astronauts orbiting the Earth are "up there where there is no gravity." If the Moon is in the gravity well of the Earth at 384,500 km, certainly the space station and any satellites orbiting the Earth at only 320 km (200 miles) are too—that they are going forward just as fast as they are falling to the Earth. So everything appears weightless.

Which brings up another misconception people have: If everything is weightless, then could we pick up an ocean liner? No, the ocean liner would pick you up, since everything still has its original mass, so attempting to pick up something more massive than you would only move you, not the ocean liner.

I remember when men first went to the Moon in 1969, one of the newspapers heralded that "now we have conquered space." This was like Balboa dipping his toe into the Pacific Ocean in Panama and claiming all the lands that bordered it, which would have included everything from Alaska to Antarctica, China, the Philippines, Japan, and Antarctica. As you will see, going to the Moon was only the tiniest step. Unlike the Universe, human arrogance really *is* infinite.

Once Galileo turned the telescope to the heavens, one of the most important—though usually ignored—discoveries he made was that the Moon was a "planet-like" body. Astronomy texts always concentrate on the fact that Galileo discovered the four moons orbiting Jupiter (which proved that not everything went around the Earth) and the phases of Venus. Both are vitally important, but I was always most impressed by the fact that a real paradigm shift took place when he discovered that the Moon that he saw in the telescope for the first time (described in his short treatise *Starry Messenger* in 1609) was not a perfect crystal celestial sphere, but was another world, the implication being that the Moon and by extension the other planets were other worlds "like the Earth." It had mountains, seas, and land like the Earth. Of course, now we know that the Moon is very different in some ways—no seas of water, only lava flows; mountains created not by plate tectonics but by huge asteroid impacts billions of years ago; and land that is not organic soil but a dry regolith (also known as Moon dirt) of dust and boulders—but still a world that could be seen in detail in even the smallest telescopes, and of course someday visited.

The first thing a lot of budding amateur astronomers look at when they get a telescope is the Moon, but they then quickly move on to other, more exotic, objects. This is a mistake, because I have found that the Moon is a very rewarding subject. The more you look at it, the more you can see. Even in my small (60 mm) Unitron, I have seen not only two-mile-diameter craters, but wrinkle ridges, rills, volcanic domes, and numerous other sights.

I made a copy of Galileo's telescope, and it is really amazing just what he figured out about the

Universe using a telescope that was far inferior to even the worst Kmart telescope you can buy today.

By the way, don't waste your money on one of those Christmas-time department store telescopes; it's not that the main lenses are so bad, but the eyepieces are terrible, and the mounts are useless. If you can't hold the object still, you can't see anything. The Earth turns at about 1,600 kph (1,000 mph) at the Equator, so objects cross the field of a telescope very fast. Also, the higher the magnification, the smaller the field of view. If you magnify the Moon fifty times, you also magnify the motion fifty times through the field of view of the telescope. Whenever I shut the motor drive off on my telescope, most people's reaction to things sailing through its field is that the scope must be moving, or the object must be moving. Finally, they figure out that the Earth is moving. Remember, you feel accelerations, not speed. This is why you can get out of your seat on an airplane and walk to the front of the airplane, even though you can't walk 502 mph. It is much better to get a good pair of binoculars and put them on a steady tripod.

Oh—and no, you can't see the flag left on the Moon by the Apollo astronauts. Because of the blurring of the Earth's atmosphere, the smallest thing you can resolve at the distance of the Moon, 1 second of arc, is about a mile. As far as I know, no one has taken anything that large to the Moon yet.

One surprising fact about looking at the Moon is that large telescopes don't show much more than backyard scopes. I remember when I was in graduate school and first got the chance to turn a large professional telescope towards the Moon—the twenty-inch refractor at Wesleyan—I was somewhat disappointed. It didn't show much more detail than my little Unitron. The reason for this is that most of the objects in which astronomers are interested in imaging are very faint, so large telescopes are needed to gather as much light as possible. This is not the case with the Moon. Although the Full Moon is some 400,000 times dimmer than the Sun, it is still very bright, because it is so close. Actually, it reflects only about 7 percent of the sunlight that hits it, just like an asphalt driveway. The issue with seeing small details on the Moon is the quality of the optics of the telescope and how steady the Earth's atmosphere is.

Photo 4: The Moon through a small telescope
(*photo by author*)

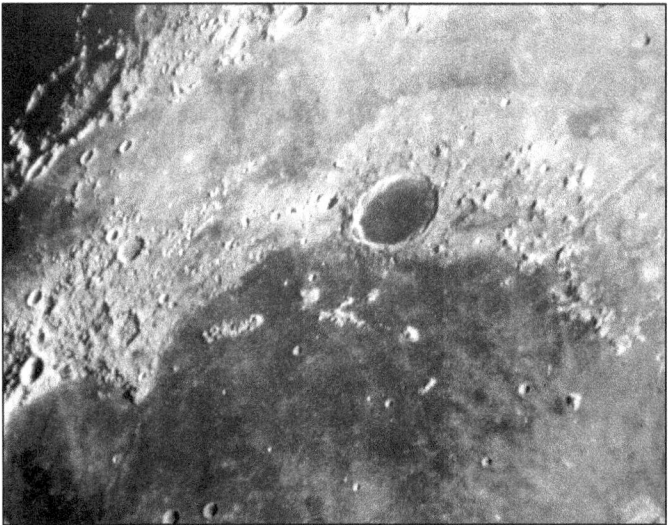

Photo 5: The Moon through a large telescope
(*photo by author*)

CHAPTER 3

THE SOLAR SYSTEM

A s the telescope developed in the seventeenth and eighteenth centuries, it was revealed that the five wandering stars or planets could be seen as disks, not just pinpoints of light. This is not the case with the stars—even the largest telescopes, which can see fainter and fainter stars, are still unable to see them as disks. For this to be possible, the stars would have to be either really big or really close. A larger telescope will show more stars but not their sizes. A few of the largest stars' disks have been resolved using a machine called an interferometer, a complicated device that can add up individual images—you can use this word to impress your friends.

This is not true for the planets. As telescopes developed larger objectives and longer focal lengths, instruments showed the planets as measurable disks. But how far away were they? Once again—just like the Moon—if we knew their actual diameters (small d) or their actual distances (big D), we could figure out how far away they were. Of course, in the seventeenth century, we knew neither. What we did figure out is their apparent size in the sky (small a), a combination of the real size (diameter d and distance D). But how to measure it?

A spider shows us the way. The story is that William Gascoigne (1612–1644) went out one night to look through his telescope, and serendipitously a spider had built a web right at the focal plane of his telescope, so it was in focus. This gave Gascoigne the idea that a wire or hair could be used to point the telescope more accurately, and he proceeded to build instruments to do just that. Unfortunately, he was killed at the battle of Marsden Moor during the First English Civil War at the age of thirty-two, so he could not follow up with observations. Fortunately, his papers—the ones that survived—eventually found their way to John Flamsteed, the Astronomer Royal.

Johannes Kepler had figured out a way to calculate the size of the Solar System by developing his three laws of motion, the last of which says that the period (P) in years squared is proportional to the distance (D) cubed, or:

$$P^2 = D^3$$

Now, the periods are easy to figure out; just go outside and track Mars or Jupiter until it returns to the same spot in the sky—in Mars's case, 1.88 years, and in Jupiter's case, 11.86 years.

So the distance to Mars is:

$P = 1.88^2 = D^3$
$3.53 = D^3$
$D = \sqrt[3]{3.53}$
$D = 1.52$

And for Jupiter:

$P = 11.882 = D^3$
$141 = D^3$
$D = \sqrt[3]{141}$
$D = 5.2$

But 1.52 and 5.2 *what*s? Well, it's Astronomical Units (AUs), the distance from the Earth to the Sun. So Mars is 1.52 times the Earth-Sun distance, and Jupiter is 5.2 times the Earth-Sun distance. The problem: We didn't know what one AU was, and there was no way of knowing without figuring out the actual distance to at least one planet. Then the AU could be calibrated.

Back to the measuring device, or, as it's now called, a reticle. Everything in the sky is so far away and so very tiny in apparent size that it is usually more convenient to measure size in seconds of arc. As noted earlier, 1 degree equals 60 minutes, and each minute equals 60 seconds, so one whole circle equals 360° or 21,600 min. or 1,296,000 seconds of arc—not of time, but of size. To give you an idea of sizes in the sky, look at the full Moon—it is one-half a degree, or 30 minutes, or 1,800 seconds, in apparent size. When Venus, which is about the same size as the Earth, is between us and the Sun (about 33 million km or 21,000,000 miles away), it can get up to 68 seconds of arc—a little over a minute. Meanwhile, even though it is much further away, Jupiter is so much bigger that it too can get to about the same apparent size, 50 seconds, or about a minute.

Modern-day reticles, which are placed at the focal point of telescope eyepieces, use etched glasses or moving wires instead of spider webs or hairs, but the principle is the same:

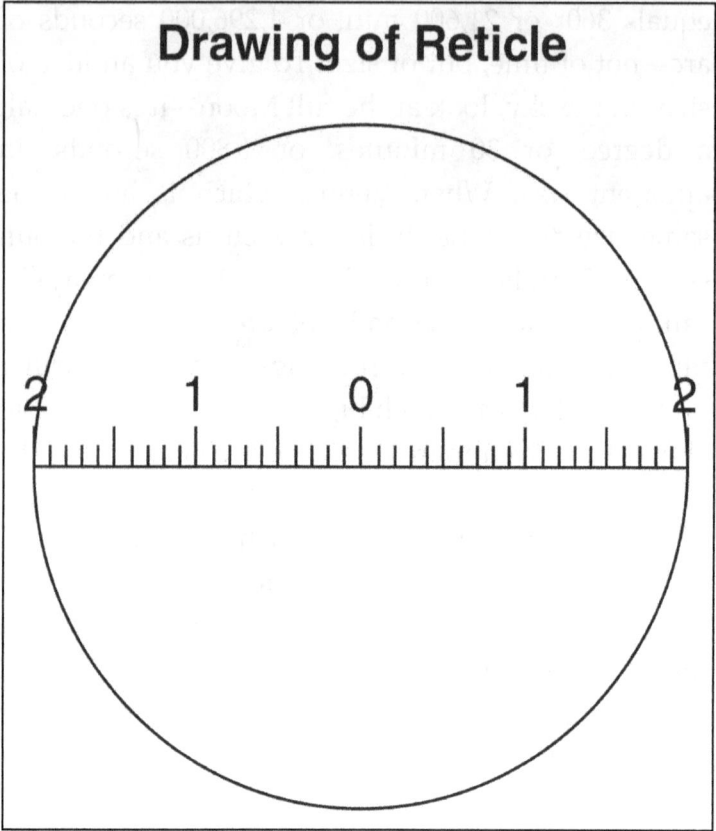

Diagram 9: Drawing of Reticle
(Each division is <u>one-tenth</u> of a millimeter)

Based on the known focal length of your particular telescope, once it is calibrated, either the actual size (*d*) or distance (*D*) can be readily calculated. In practice, how this works is based on the simple principle that the farther away something is, the smaller its apparent size. Go out in a parking lot and look at your car (or anybody else's car) from a good distance away—how far away is it? Well, unlike the planets, you could measure it using a reel-type tape measure, or you could get a model of the car and hold it in front of your eyes at the distance where the model and the real car appear to be the same size. For example, you hold the model away and measure the distance from your eyes and it is 22.5 cm (9 inches); look underneath the model and it will say, for example, 1:64, which means that it is one sixty-fourth the size of the real car.

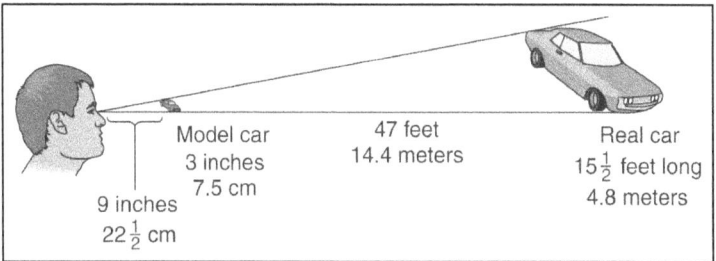

Diagram 10: A person holding a model car, with a real car in the distance

So the real car is sixty-four times bigger than the model; you measure the model and it is 7.5 cm (3 inches) long, so the real car at sixty-four times larger is 480 cm/4.8 meters (187 inches/15.5 feet) long. The model car would have to be sixty-four times farther away than 22.5 cm to show the real comparative sizes. Then 22.5 cm times sixty-four equals 1,440 cm, or 14.4 meters (47 feet)—that's how far away the real car is. Of course, this only works because I know how long the real car is by multiplying 7.5 cm times sixty-four. In the case of the planets, astronomers didn't know the real sizes *or* their distances, but if they knew one of the two measurements, the other would be simple to calculate. Using what is called the small-angle equation, let a small *a* be the planet's (or anything else's) apparent size in the sky in seconds of arc, with a small *d* as the actual diameter or size, and *D* the distance. Then the relationship is simply:

$$Apparent\ size\ a = \frac{d \times 206{,}265}{D}$$

$$Distance\ (D) = \frac{206{,}265 \times d}{a}$$

$$diameter\ (d) = \frac{aD}{206{,}265}$$

Where does the magic number 206,265 come from? Well, if you take the length of the radius of any circle big or small and lay it along the circumference, it will always subtend (or cover) 57.2958 degrees or, times 60 minutes per degree =

3,437.75 minutes, then times 60 again equals 206,265 seconds of arc. This is called a radian.

So, 57 degrees × 60 min/degree × 60 seconds/minute = 206,265 seconds of arc in a radian.

Armed with a reticle and better telescopes, astronomers were ready to attack the problem of the size of the Solar System. Go get a pencil and hold it up about 15 cm (6 inches) from your face. Close first your left eye, then open it, and then close your right eye. See how it jumps back and forth as you look with either your right or left eye? Now hold the pencil at arm's length and do it again. It still jumps back and forth, but you will notice not as much. This is because your baseline—how far apart your eyes are—is not very big. Stereoscopic vision (having both eyes in front of your head) works for close objects; it will keep you from falling down the stairs—but it fails after only a few hundred yards. We estimate how far away distant objects are by some knowledge of how big they really are. Unconsciously, if you're looking at a house that subtends 30 degrees of your vision and one that subtends 5 degrees, you automatically deduce that the one at 5 degrees must be farther away. Of course, there could be a really tiny house right next door to a really big house, but that is unlikely.

In the seventeenth century, astronomers realized that this principle—called parallax—could be applied to computing the distances to the planets, but they would need a lot bigger baseline than the separation between human eyes. As they were confined to the Earth, the idea was to get as far apart on the Earth as they could so that the baseline would be as large as possible. Giovanni Cassini, director of the Paris Observatory at the time, sent a colleague, Jean Richer, to French Guiana in South America in 1672. The idea was to measure the parallax of Mars simultaneously from Paris and South America.

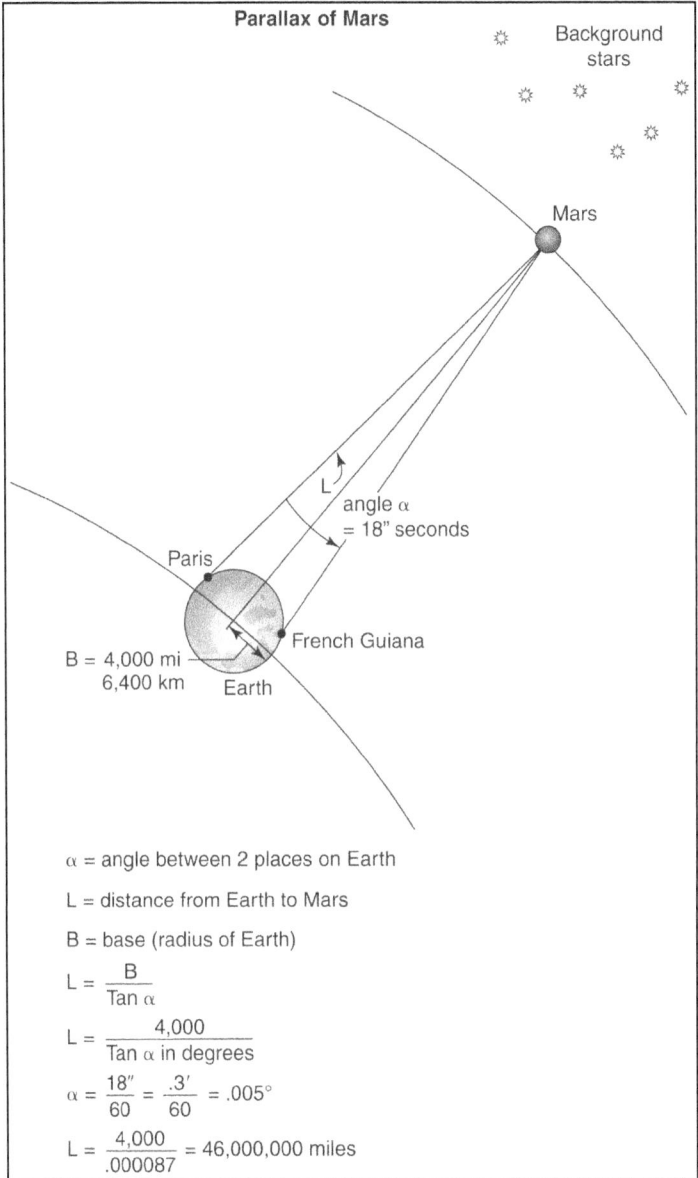

Parallax of Mars

Background stars

Mars

L

angle α = 18" seconds

Paris

B = 4,000 mi
6,400 km

French Guiana

Earth

α = angle between 2 places on Earth

L = distance from Earth to Mars

B = base (radius of Earth)

$$L = \frac{B}{Tan\ \alpha}$$

$$L = \frac{4,000}{Tan\ \alpha\ in\ degrees}$$

$$\alpha = \frac{18''}{60} = \frac{.3'}{60} = .005°$$

$$L = \frac{4,000}{.000087} = 46,000,000\ miles$$

Diagram 11: Parallax of Mars

In Diagram 11:

$$L = \frac{4,000}{\tan a} = \frac{4,000}{\tan .005} = \frac{4,000}{.000087} = 46,000,000 \text{ miles}$$

So Mars is 74,000,000 km from the Earth. From Kepler's laws we know that, proportionately, it is about 1.5 AUs from the Sun, so 2 × 74,000,000 equals 148,000,000 km for one AU—pretty close to today's figure of 149,597,871 km. And Jupiter, which is 5.28 AUs, is simply 5.2 × 150,000,000 km (780,000,000 km) from the Sun, almost half a billion miles! Nowadays, if you have a small telescope and a reticle, you can measure the distances or sizes of the planets from your backyard. Jupiter is 142,000 km in diameter. The focal length of my very small (60 mm) telescope is 900 mm, so the plate scale—the angular field of view of a particular telescope—is 206,265 divided by 900 mm, which equals 229 seconds of arc per millimeter; therefore every millimeter on the reticle equals 229 seconds of arc in the sky. You put Jupiter in the telescope and place it so that the scale divides right across Jupiter's equator like so:

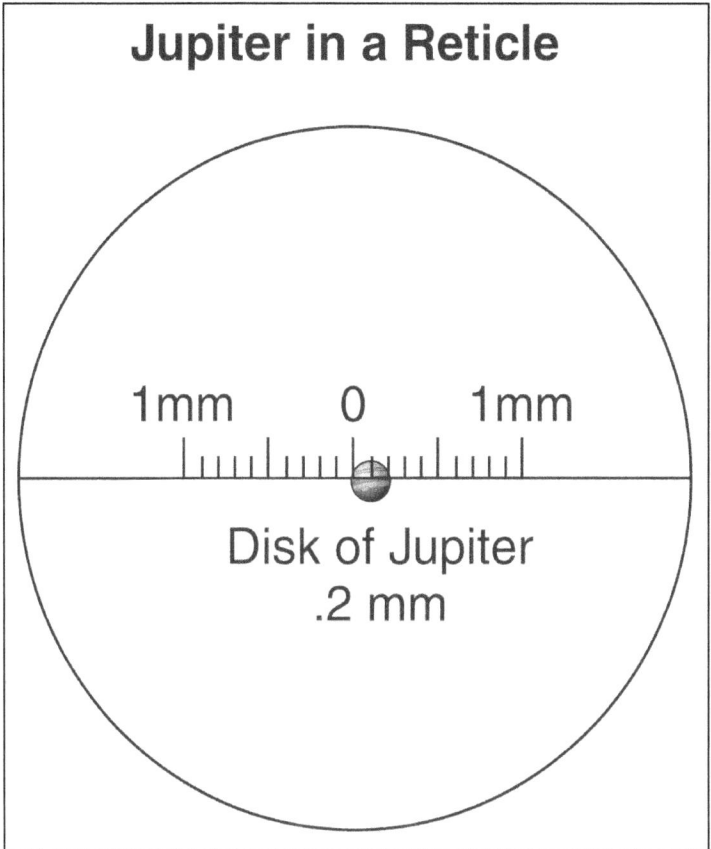

Diagram 12: Jupiter in a Reticle

Read that the planet covers two marks, or .2 of a millimeter. This is harder to do than it sounds (a motor drive is necessary to keep the planet in the reticle because of the Earth's motion), and Cassini derived even an approximately correct answer for the AU—86,000,000 miles—only through a series of mutually self-canceling errors. Using the same principle, by observing transits of Venus across the Sun in the eighteenth century, Jérôme Lalande was able to improve the distance to about 94,000,000 miles in 1761 and 1769. In fact, the last time parallax was used to determine the AU was 1941; now the Astronomical Unit has been refined by radar and telemetry to exactly 149,597,870.7 km, which is 92,955,807.273 miles. To calculate the distance to Jupiter:

On reticle scale the distance large D equals:

$$D = \frac{206,265 \times d}{\text{seconds per mm} \times \text{measurement in mm}}$$

$$D = \frac{206,265 \times 88,600 \text{ miles}}{229 \times .2 = 45.8}$$

$$D = \frac{1.8275 \times 10^{10}}{45.8}$$
$$= 400,000,000 \text{ miles}$$

In metric it would be:

$$D = \frac{206,265 \times 142,000 \text{ km}}{229 \times .2}$$
$$= \frac{2.927 \times 10^{10}}{45.8} \text{ km} = 640,000,000 \text{ km}$$

I have done this on numerous occasions, and I find that, with practice, I can usually get about 95 percent to 98 percent accuracy. Of course nowadays we don't have to do any of this; astronomers just bounce radar beams off the planets and get the distance. Blast a radar or laser beam to the Moon, and returns (very little of it) in 2.6 seconds after traveling to the Moon and back. Dividing 2.6 seconds by two is 1.3 seconds, and the speed of light is 300,000 kps (186,000 mps), so 300,000 kps times 1.3 is 390,000 km (241,800 miles), the distance to the Moon. At the speed of light it's about four minutes to Mars when it's at the closest, thirty-five minutes to Jupiter, one hour and ten minutes to Saturn, and four hours to Pluto.

How can we envision the distance to Jupiter? I can imagine a car trip of 160 km (100 miles) at 80 kph (50 mph), or, in an airplane, a trip an order of magnitude faster (800 kph) across the Atlantic Ocean, but how do you imagine a journey of 640,000,000 kilometers? Let's make a model. Almost everyone has a model of the Earth around the house left over from when you were interested in geography as a kid. They're usually about 30 cm (12 inches) in size. Paint it yellow (actually white, because if the Sun were yellow, snow would look yellow), so that the Sun will now be a 30 cm globe. The Sun is actually 1.4 million km (865,000 miles) in diameter, so 1 mm equals 4,700 km (2,900 miles) in the model. To represent the distance to the Earth from the Sun, 150,000,000 km (1 AU) divided by 4,700 km/mm = 32,000 mm or 32 meters (96 feet) away. The Earth is about a hundredth the size of the Sun, so 30 cm (300 mm) divided by 100 = 3 mm, the size of the head of a pushpin; the Moon would be about 80 mm away and one-quarter the size of the Earth, an even smaller pinhead. Tiny as the Moon is compared to the Sun, the reason that the Moon can completely cover the Sun is that it has moved away from the Earth in the last 4 billion years about 35 mm per year, so right now it is the same apparent size as the Sun in the sky. As noted earlier, the Moon is four hundred times smaller than the Sun, but also four hundred times closer—just like our car model was sixty-four times smaller and sixty-four times closer.

How about the rest of the Solar System? Mars would be one-half the size of the Earth, and the closest it would ever get would be 16 meters (about 50 feet). Jupiter, the largest planet, is about eleven times the diameter of the Earth, so about 30 mm (a little over an inch) in diameter 136 meters (400 feet) away. Saturn, a little smaller than Jupiter, would be 300 meters or 900 feet away. The last two planets (now that Pluto has been demoted), Uranus and Neptune, at 19 and 29 AUs respectively, would be

600 and 925 meters (1,800 feet and 2,800 feet) from the Earth—that's a third of a mile and half a mile away to the last little pebbles in the Solar System. Going the other way—towards the Sun—it is just as empty; only Venus, about the same size as the Earth, and Mercury (only a little bigger than the Moon) are between the Sun and us.

This is why I don't worry too much about an errant asteroid running into us. The Earth is a really tiny and really fast-moving target, moving at 10,500 kph (66,000 mph) around the Sun.

And there's another reason not to worry. It's your infinitesimally short life span. A large asteroid hits the Earth about once every 100,000 years. The chances of this happening during your short stay on the Earth are really remote. It could happen, but lots of other things are much more threatening—like crazy people armed with nuclear bombs or a worldwide plague. Even though there are billions of people, I know of only one case where someone was actually hit by a meteor—a woman in Alabama in 1954 that lived, appropriately enough, across from the Comet Drive-In. As I recall from looking at the pictures, she was a somewhat larger target than most of us.

Think about it—a 30 cm globe inside a one-and-a-half-kilometer sphere with nothing else except a few tiny pebbles and pinheads orbiting around it. Mass-wise, it is even more striking—the Solar System is mostly the Sun and Jupiter and leftover junk. Consider it like this: If the Solar System were a small town of 3,000 people, 2,998 of them would be the Sun, one could be Jupiter, and the one person left over would represent all the other planets, including the Earth, Saturn, etc., plus all the moons, asteroids, and comets.

CHAPTER 4

THE STARS

S tars are basically simple creatures, not anywhere near as complicated as grasshoppers. Two forces are at play in the life of a star—gravity trying to crush it out of existence and gas pressure from the nuclear reactions in the core trying to blow it up. When these two forces are in balance, you have a star. This is called hydrostatic equilibrium. A star's whole existence depends upon one parameter—how much mass it has when it is born. Most stars are made of the same stuff, mostly hydrogen and helium.

Hundreds of years ago, when it was first realized that the Sun was a star, the assumption was that all the stars were like the Sun, so it was thought that the dimmer stars must be farther away. One of the arguments against the Copernican heliocentric theory—that the Sun was the center of the universe, motionless, with the Earth and the other planets revolving around it in more or less circular paths—was this: If the stars were at different distances, using the parallax method, why didn't the closer ones shift back and forth compared to the more distant ones as the Earth went around the Sun? Obviously the stars are too far away to use the diameter of the Earth as a baseline, but without leaving the Earth, we have a much bigger baseline—the orbit of the Earth itself.

Once the Astronomical Unit was fairly well established—by the late eighteenth and early nineteenth centuries—at around 150,000,000 km or 93,000,000 miles (one AU), looking at a suspected nearby star in June and then again in December should show some parallax. The first problem was which stars to choose—one would think that the brighter stars must be the closest—so in the mid-nineteenth century, two astronomers, Thomas Henderson in South Africa and Friedrich Struve in what is now Estonia, tried to measure the distances to Alpha Centauri and Vega, two of the brightest stars. They did succeed, but they were beaten by Friedrich Bessel, a Prussian who decided to measure the parallax not of a bright star, but of a star across the sky rather quickly—61 Cygni.

As mentioned earlier during our discussion of the planisphere, the stars do move across the sky. It's just that they are so far away that centuries must pass before we notice it. In fact, it was Edmond Halley, of comet fame, that first noticed this "proper motion," as it is called by comparing contemporary star maps with the ones that were in Ptolemy's *Almagest* in ancient times. The closer ones appear to move faster for the same reason that when you are driving down the road, fence posts appear to go by quicker than a distant water tower, even though they are both fixed to the ground. It was very difficult to measure the distance to the stars because of one simple fact: The stars—*all* of the stars except for the Sun—are far away. *Very* far away. In fact, so far away that we can't even use miles, kilometers, or AUs to describe their distances.

Here's an example: Look at a roadmap of the United States. It is 250 miles from Cleveland to Cincinnati—that's 1,320,000 feet or 158,400,000 inches and, at 25.4 mm per inch, 402,336,000 mm. Now look at the distance from Cleveland to Miami, Florida—1,241 miles or 1,997,195,904 mm. I can readily envision the difference between 250 miles and 1,241 miles, but the difference between almost half a billion and 2 billion escapes me.

Similarly, in Europe it is 2,880 km from Paris to Moscow, or 2,880,000,000 mm. It's all just big numbers. That's why we don't use millimeters to drive between cities on the Earth. In a like manner, we don't use miles or kilometers when talking about distances to the stars. For example, it's 150,000,000,000,000 miles (240,300,000,000,000 km) to Vega, a nearby star. You can readily see that astronomy books would be mostly filled with zeroes from all the big numbers. A more convenient measurement is based on the distance from the Earth to the Sun, one AU, and if a background star should shift it back and forth by one second of arc in a year, it would be one parallax second away, which is referred to as 1 parsec (pc).

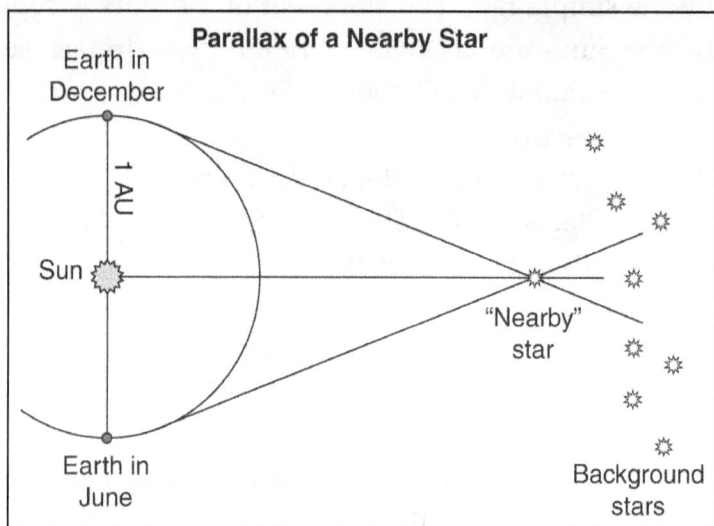

Diagram 13: Parallax of a "Nearby" Star

Well, how much of a shift is this? Hold the pencil we used to demonstrate parallax in front of your eyes at arm's length again (about 20 inches or 500 mm). By alternately opening and closing your right eye, then your left eye, the pencil will shift about 5 degrees, 300 minutes, or 18,000 seconds. Remember, as you move the pencil farther and farther away, it shifts less and less—in order to make it shift one second of arc, you would have to move it 18,000 times farther away (18,000 × 20 inches = 5.5 miles or 9 kilometers)!

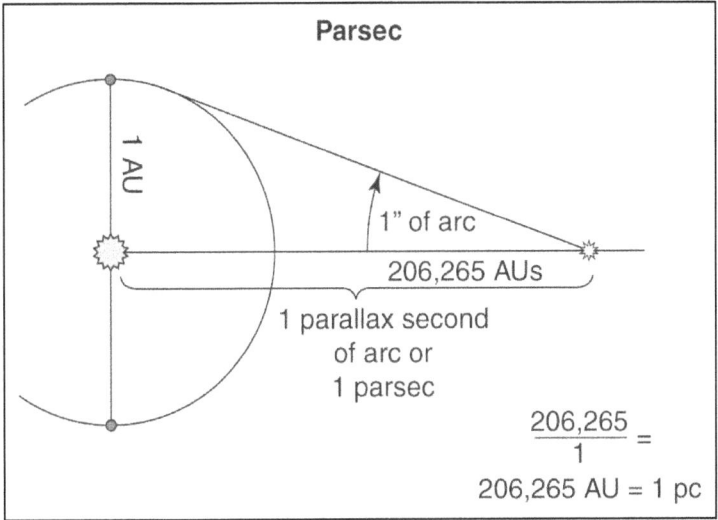

Parsec

1 AU

1" of arc

206,265 AUs

1 parallax second
of arc or
1 parsec

$$\frac{206,265}{1} =$$

206,265 AU = 1 pc

Diagram 14: Parallax Shift of 1 Parsec

Obviously, the baseline of the distances between your eyes is too short to notice this. And in like manner, the telescopes and reticles available to early nineteenth century astronomers were not accurate enough to detect so small a shift either. To add to the problem, *no* star is even this close. They are all farther away than 1 parsec, so astronomers were trying to measure tenths of a second of arc for even the closest stars. Using what was arguably the best telescope available, the Dorpat 9.5-inch Refractor, Friedrich Bessel managed to measure the parallax of 61 Cygni at .314 seconds of arc (present day value .285), so he had about 9 percent error. This translates into:

$$.285 = 3.5 \; pc \times 3.26 \; ly/pc = 11.4 \; light\text{-}years$$

Most popular astronomy books use light-years as a distance unit, and there are 3.26 light-years in a parsec, so 3.5 pc times 3.26 light-years per parsec equals 11.4 light-years. Try to imagine this—it takes light going 1 billion kph (669,000,000 mph) 8.3 minutes to get to the Earth from the Sun, and 11.4 years to get to 61 Cygni. That's 11.4 years times 365.25 days in a year times 24 hours in a day times 60 minutes in an hour, or 5,995,944 light minutes to 61 Cygni; 5,995,944 light minutes divided by 8.3 light minutes equals 722,403 AUs, or Earth-Sun distances, to a relatively nearby star system. By the end of the nineteenth century, astronomers had fairly good parallaxes to about sixty stars. The very closest turned out to be the Alpha Centauri system, at 1.3 pc, or 4.3 light-years away. Actually, it's a triple star system consisting of a star like the Sun, one slighter dimmer, and one very dim red dwarf, called Proxima Centauri, because it's the closest of all at 4.24 light-years from us.

Let's try to construct a model of how far away Alpha Centauri is: If the Sun were a tennis ball 63 mm (2.5 inches) in diameter, this would represent the actual diameter of the Sun, which is 1.4 million kilometers (865,000 miles). In this model, 1.4 million kilometers (865,000 miles) divided by 63 mm equals 22,000 km/mm, so the Earth at 150,000,000 km from the Sun would be 150,000,000 km divided by 22,000 km/mm, which equals 6.8 meters away, or 20 feet. If we represent the Sun by a tennis ball, then the Earth would be one hundredth of that size, .5 mm, and the Moon even tinier at 12 mm (half an inch) from the Earth. Recall that some newspapers proclaimed that man had "conquered space" after the 1969 Moon landing. It took us three days to go 240,000 miles, or half an inch. The last planet, Neptune, at 30 AUs, would be 200 meters (22 feet times 30 or 672 feet away, a little over two football fields). Alpha Centauri is 1.31 pc (4.2 light-years) away; that's 1.31 × 206,265 AUs per parsec = 270,207 AUs × 6.8 meters/AU = 1,837,407 meters, or, rounding off, some 1,800 kilometers (about 1,100 miles) proportionately—and there's absolutely nothing in between the Solar System and it except for some stray asteroids and comets.

Just imagine it this way—one tennis ball, and then nothing between Cleveland, Ohio, and New Orleans, Louisiana, or Hamburg, Germany, and Madrid, Spain. How long would it take us to reach Alpha Centauri? Well, the Voyager 1, having left the

Solar System after being launched in 1977, is traveling at 62,000 kph (39,000 mph) which sounds pretty fast; it's about 128 AUs away now, about 19,000,000,000 km (12,000,000,000 miles) or about a kilometer (half a mile) in our model. The Voyager is traveling at 39,000 mph. That's 669,000,000 mph divided by 39,000 mph—1/17,000 the speed of light. The distance to Alpha Centauri, 4.3 light-years times 17,000, is over 70,000 years; and that's to get to even the nearest star.

When astronomers started creating a census of the nearby stars, they found out some interesting things. Not only were they very far apart, but they also varied enormously in their luminosity, what astronomers call absolute magnitude. Apparent magnitude, on the other hand, is just what it sounds like—how bright something appears from the Earth. This magnitude scale is based on a nonintuitive and somewhat arbitrary scale set up in the nineteenth century stating that, by definition, sixth magnitude stars—the dimmest stars you can see without a telescope—are one hundred times dimmer than first magnitude stars; that's five steps, not six, since it started with one, so each step or whole magnitude is by definition the fifth root of 100, or 2.512 times dimmer than the one before. The larger the number, the dimmer the star, so eighth magnitude stars are 2.512 to the seventh power times brighter than fifteenth magnitude stars, or 631 times brighter, and tenth magnitude stars are 2.512 to the third power times dimmer than seventh magnitude stars, or sixteen times.

To add to the confusion, apparent magnitudes are always indicated by a lower case *m* and absolute magnitudes by a capital *M*. The Sun at 10 pc would have an absolute magnitude of +4.8, just barely visible to the naked eye. It's important to put the plus sign on the number, because when stars were compared and exact magnitudes derived, it was discovered that some stars, like Sirius and Canopus, were brighter than the standard first magnitude star Vega; the obvious thing to do therefore was to make them zeroes or minus magnitudes, so Sirius's apparent magnitude is −1.44 and its absolute magnitude is +1.45.

The bottom line is that absolute magnitudes are simply the apparent magnitude a star would have at 10 pc. There is nothing magical about the 10 parsecs distance, except that this is a way to imagine mathematically how bright they would all be if they were all at the same distance; the distance chosen was 10 pc because the parallax would be one-tenth of a second of arc.

The other thing astromomers noticed, once spectrum analysis was discovered, was that the stars vary in temperature, from about 2,500 degrees

Kelvin to over 40,000 degrees Kelvin, and that this is reflected by their color. When you turn your electric stove on high, the burner turns red; if you could keep turning it up it would turn yellow, then white, and finally blue—though it would have melted long before this. So the red stars are the coolest and the blue ones are the hottest.

The distances, temperatures, and luminosities were slowly collected. At the beginning of the twentieth century, two astronomers, Ejnar Hertzsprung and Henry Norris Russell, collected all this information and created a diagram that has proven to be one of the most important tools of modern astronomy—the H-R diagram. Two scales on the x axis of this diagram delineate the temperature, or spectral type, noted by the sequence of letters OBAFGKM from hottest to coolest (and within each one further refined by the numbers 0–9). The y axis is the absolute magnitude, or luminosity, in solar units, the Sun being one solar luminosity. So the Sun is a G2V star, the V denoting the fact that it is a main-sequence dwarf. (More on this in a few pages.)

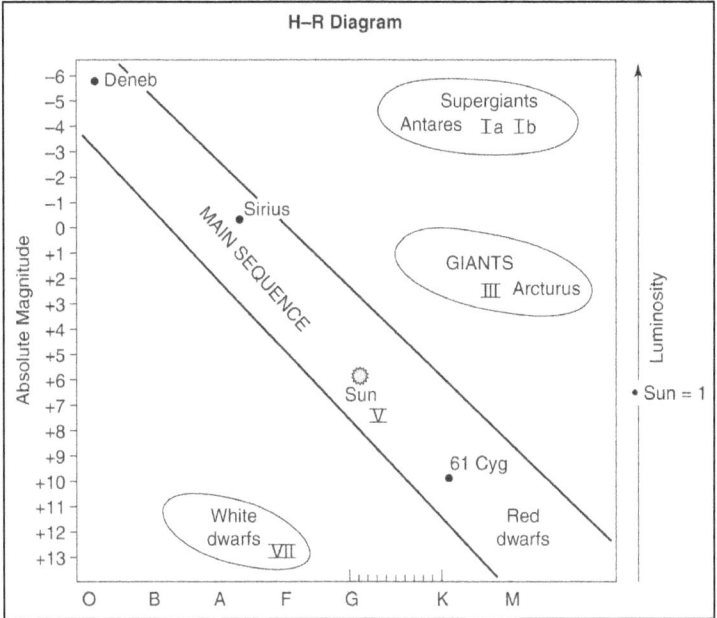

Diagram 15: H-R Diagram

Remember that at the beginning of the twentieth century, the sample of stars with good distances and spectra (their color or temperature) was fairly small, only about sixty or so. Think of it this way: If aliens came to Earth and landed in Oslo, Norway, they would probably conclude that the average person has blonde hair and blue eyes, but they would draw a very different conclusion if they landed in Tokyo, Japan. They would need a much better sample; if they had landed in New York City, or if they would have surveyed an entire continent like Europe, they would have gotten a much better sample of all types of people.

So the question is, is the Solar System in a very typical part of our galaxy? Do the ones that are nearby represent all types of stars?

Well, there are no red supergiants, massive blue stars, or neutron stars near us, to say nothing of the fact there are no black holes (thankfully). What Hertzsprung and Russell *did* notice was that the stars' temperatures and luminosities were not scattered randomly around the chart; the vast proportion of the stars populated a fairly wide band going from the hottest, most luminous stars in the upper left corner to the dimmest, coolest stars in the lower right.

Also, the smaller, cooler stars were much more numerous than the giant hot blue stars. The reason for this is the same reason that there are more ants than elephants (or, if you drop a vase on the floor, you usually get a lot more small fragments than big pieces). There are 10,000,000,000,000,000 ants on our planet and only 7,000,000,000 people— that's about 1.5 million ants per person—and in biomass they still outweigh us by a lot. Actually, if aliens ever do come to the Earth, they would no doubt call it the "Planet of the Insects."

Charles Darwin once remarked that God must have loved beetles the most, since he made more species of them than all other animals combined.

		RA (2000) Dec		π	επτ	D	μ	θ	Vrad			
Name		h m	° '	mas	mas	ly	mas/y	°	km/s	Sp.	mv	Mv
Sun	+8P									G2V	−26.72	4.83
1 Proxima Cen	(C)	14 30	−62 41	768.85	0.29	4.24	3853	281.5	−16	M5.0V	11.05	15.48
α Cen	A	14 40	−60 50	747.23	1.17	4.36	3710	277.5	−26	G2V	0.01	4.38
	B	14 40	−60 50				3724	284.8	−18	K0.0V	1.34	5.71
2 Bernard's		17 58	+04 42	545.51	0.29	5.98	10358	355.6	−111	M3.5V	9.57	13.25
3 Wolf 359		10 56	+07 01	419.10	2.10	7.78	4696	234.6	+19	M5.5V	13.53	16.64
4 Lalande 21185		11 03	+35 58	393.25	0.57	8.29	4802	186.9	−85	M2.0V	7.47	10.33
5 Sirius	A	06 45	−16 43	380.02	1.28	8.58	1339	204.1	−9	A1V	−1.43	1.47
	B	06 45	−16 43							DA2	8.44	11.34
6 BL Cet	(A)	01 39	−17 57	373.70	2.70	8.73	3368	080.4	+29	M5.5V	12.61	15.47
UV Cet	(B)	01 39	−17 57						+32	M6.0V	13.06	15.92
7 Ross 154		18 50	−23 50	337.22	1.97	9.67	666	106.8	−10	M3.5V	10.44	13.08
8 Ross 248		23 42	+44 11	316.37	0.55	10.31	1617	177.0	−78	M5.5V	12.29	14.79
9 ε Eri	+1P	03 33	−09 27	311.22	0.09	10.48	977	271.1	+16	K2.0V	3.73	6.20
10 CD−36 15693		23 06	−35 51	305.08	0.70	10.69	6896	078.9	+9	M1.0V	7.34	9.76
11 Ross 128		11 48	+00 48	298.14	1.37	10.94	1361	153.6	−31	M4.0V	11.16	13.53
12 EZ Aqr	A	22 39	−15 18	289.50	4.40	11.27	3254	046.6	−50	M5.0VJ	13.03	15.34
	B	22 39	−15 18								13.27	15.58
	C	22 39	−15 18								15.07	17.38
13 61 Cyg	A	21 07	+38 45	286.08	0.48	11.40	5281	051.9	−66	K5.0V	5.20	7.48
	B	21 07	+38 45				5172	052.6	−64	K7.0V	6.03	8.31
14 Procyon	A	07 39	+05 13	285.08	0.64	11.44	1259	214.7	−4	F5 1V–V	0.37	2.65
	B	07 39	+05 13							DQZ	10.70	12.98
15 BD +59 1915	A	18 43	+59 38	283.08	1.46	11.49	2238	323.6	−1	M3.0V	8.90	11.17
	B	18 43	+59 38				2313	323.0	+1	M3.5V	9.69	11.96
16 GX And	(A)	00 18	+44 01	279.87	0.60	11.65	2918	081.9	+12	M1.5V	8.08	10.31
GQ And	(B)	00 18	+44 01						+11	M3.5V	11.06	13.29
17 ε Ind	A	22 03	−56 47	276.07	2.28	11.81	4704	122.7	−40	K3.0V	4.68	6.89
	B	22 04	−56 47				4823	121.1		T1.0V		
	C	22 04	−56 46							T6.0V		
18 DX Can		08 30	+26 47	275.80	3.00	11.83	1290	242.2	−5	M6.0V	14.90	17.10
19 τ Cet		01 44	−15 56	273.97	0.17	11.91	1922	296.4	−17	G8.5V	3.49	5.68
20 GJ 1061		03 36	−44 31	272.01	1.30	11.99	831	118.8	−20	M5.0V	13.09	15.26
21 YZ Cet		01 13	−17 00	269.08	2.99	12.12	1372	061.9	+28	M4.0V	12.10	14.25
22 Luyten's		07 27	+05 14	266.23	0.66	12.25	3738	171.2	+18	M3.5V	9.85	11.98
23 SCR 1845−6357	A	18 45	−63 58	259.50	1.11	12.57	2558	074.7		M8.5V	17.40	19.47
	B	18 45	−63 58							T6.0V		
24 SO 0251+1652		02 53	+16 53	259.41	0.89	12.57	5050	137.9		M6.5V	15.14	17.21
25 Kapteyn's		05 12	−45 01	255.67	0.91	12.76	8670	131.4	+245	M2.0V1	8.85	10.89
26 AX Mic		21 17	−38 52	253.44	0.80	12.87	3455	250.6	+28	K9.0V	6.67	8.69
27 DENIS 1048−3956		10 48	−39 56	248.53	1.18	13.12	1530	229.2		M8.5V	17.39	19.37
28 Kruger 60	A	22 28	+57 42	248.06	1.39	13.15	990	241.6	−34	M3.0V	9.79	11.76
	B	22 28	+57 41							M4.0V	11.41	13.38

TABLE OF NEAREST STARS

Table 1: Nearest stars, listed in the Royal Astronomical Society of Canada's *Observer's Handbook 2013*, p. 287

So when the first H-R diagrams were created based on the stars in which we could figure out the distances—i.e., the nearby ones—the types of stars were heavily weighted towards the red dwarfs.

In the list of nearby stars in Table 1, arbitrarily set out to 4 parsecs (13 light-years), there are two hot A-type stars, Sirius and Procyon; three Sun-like stars, including the Sun, Alpha Centauri A, and Tau Ceti; five K-type orange dwarfs; two hot little white dwarfs; twenty-seven red dwarfs; and a few brown dwarfs that are almost stars. That's forty stars, twenty of which are in multiple star systems. Notice that 69 percent of them are red dwarfs. If our little corner of the Milky Way is typical, and there's no reason to believe it isn't, then most of the stars in the sky are invisible to the naked eye.

The other parameter that stars vary in is in their sizes. Remember, there are red dwarfs and red supergiants, both of about the same temperature—about 3,500 degrees Kelvin. And they all appear like points of light in the telescope. So how can astronomers tell the difference?

Once again, it is by examining their spectra. The larger a star is, the more tenuous its atmosphere is. Temperature is caused by collisions between atoms, but the *rate* of collisions depends partly on the density of the star; main-sequence stars like the Sun and red dwarfs have denser atmospheres than giants or supergiants do, hence more collisions—a given spectral line will therefore appear thicker. So, if you know the luminosity and temperature of a star, you can figure out the size in solar radii (R*). For example:

R* = (5,800/Temp of *)² × √lum of the star, which says that the radius of the star compared to the Sun is equal to the temperature of the Sun divided by the temperature of the star squared times the square root of the luminosity of the star compared to the Sun. Here are some examples.

For an M dwarf like Barnard's Star:

R = (5,800/3,100)² × √.0004 = 3.5 × .02 = .07 × R* (the radius of the Sun) × 2 = the diameter = 60,550 miles*

So that's 432,500 miles (or 692,000 km) × .07 × 2 = 60,550 miles (97,450 km)—only 7.5 times the size of the Earth.

An M-type supergiant of the same temperature but a lot higher luminosity would be:

R = (5,800/3,100)² × √20,000 = 3.5 × 141 = 494 (about 500) = 432,500,000 miles*

So 432,500 (radius of the Sun) × 500 × 2 = 430,000,000 miles (690,000,000 km), which means if it were in our Solar System, all the inner planets up to Mars would be orbiting down inside it—because it is so big that its radius is almost halfway out to the orbit of Jupiter.

The Morgan–Keenan (MKK) system is used in stellar classification. The letters O, B, A, F, G, K, and M are used to indicate a star's temperature. Astronomers add a Roman numeral to the description of its spectral class. That Roman numeral at the end tells you the star's size—*Ia*, *Ib*, and *II* are the supergiants; *III* are the giants like Arcturus; *V* are the main-sequence stars like the Sun; and *VII* are the white dwarfs, which are just as hot as main-sequence A-type stars but are much smaller. An example of this is Procyon and its white dwarf companion, Procyon B.

Procyon A is:

$R^* = (5800/6530)^2 \times \sqrt{6.93} = .788 \times 2.6 = 2.07 \times$ *size of the Sun = 1,790,550 miles*

That's about twice the size of the Sun, but its companion Procyon B (a white dwarf) is only a little hotter but much less luminous, so:

$R^* = (5800/7740)^2 \times \sqrt{.00055} = .56 \times .023 = .012$ *times the size of the Sun*

865,000 × .012 = 10,380 miles, or in metric 1,400,000 km × .012 = 16,800 km

That's only a little bigger than the Earth— 12,800 km (8,000 miles). How can we envision this? I like to think of the stars as tomatoes.

Photo 6: Stars as Tomatoes (*photo by author***)**

Like people, every tomato and every star is an individual. We can make up any one of a number of taxonomies, but real nature never reproduces exactly the same thing twice. My wife likes to grow tomatoes, and they are all pretty much look the same if you don't look too close. Most of them are sort of round and, at least for the cherry tomatoes, about the same size.

But if you look closely, every one looks subtly different. Some are oblate, like some stars. For example, Altair rotates so fast it is no longer spherical. Some have grown so close together that they have merged. Again, if some stars get inside the Roche distance, which is the minimum distance bodies can exist without falling apart—or, in some cases, actually merging into each other—they will be so distorted that they will exchange matter, the way W Ursae Majoris stars do.

And of course different species of stars, like different species of tomatoes, are of vastly different sizes. Take one of the very largest stars that we know of, VV Cephei. If it was a tomato and the Sun was represented by the size of a cherry tomato—25 mm (about an inch)—VV Cephei would be 45 meters (1,800 inches—that's 150 feet) across. And that's a big tomato! If it were to replace the Sun (the star, not the tomato) in the Solar System, it would engulf the entire Solar System, out to Saturn. Remember, though—the outer atmospheres of these infrared supergiants are so tenuous that you would have to travel millions of miles into them before you would even notice that you were inside a star.

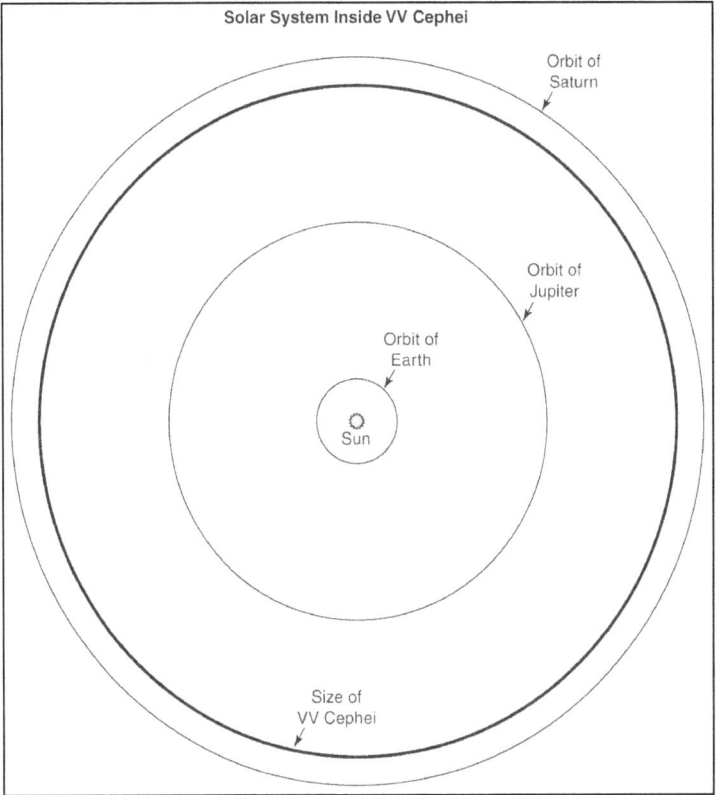

Diagram 16: Solar System Inside VV Cephei

If you go outside and look at the sky, it's a safe guess that the typical star will be an A- or B-type main-sequence star—neither too faint to see, like the red dwarfs, nor too rare, like the blue supergiants. In fact, of the ninety brightest stars, about half of them are As or Bs—Bs because they are so luminous that they can be seen over enormous distances. Most of the rest are giants or supergiants, because even though they are not hot, they are so big that their luminosity is very high. According to the Royal Astronomical Society of Canada's *Observers Guide*, the 314 brightest stars— that is, the ones that are brighter than +3.55 apparent magnitude—come to about 400 light-years distance from us, if you average them out. There are a few exceptions, but if you're looking up at the sky at night and someone asks you how far away you think that star is, a safe guess would be tens or hundreds of light-years, and certainly not thousands or millions.

Sometimes people ask me what the farthest star from Earth is that you can see with just your eyes. Certainly the easiest to find would have to be Epsilon Aurigae at 2,000 light-years, but if you can find Rho Cas at an apparent magnitude of +4.5, it is over 10,000 light-years away.

In a 4 pc sample, there are no B stars, no supergiants, and not even any regular giants—the closest red giant to us is Pollux, at 34 light-years. Even within 4 pc, we have to shrink our model down a lot; tennis balls are too big. We can no longer scale both sizes and distances. If we use them to represent stars, even very small pinheads, at one-sixteenth of an inch in diameter, are too large when we want to show the distances.

For example, if the Sun were one-sixteenth of an inch (1.5 mm) in diameter, representing 865,000 miles (1.4 million km), the Earth would be an invisibly small one-hundredth the size of a pinhead dot at 6.6 inches (17 cm) away, and the nearest star, Alpha Centauri, at 270,207 AUs away times 6.6 inches, would be 6.6 inches × 1,783,366 inches, or 28 miles (45 km) away!

So, for the sake of trying to comprehend the nearby stars, and knowing full well that pinheads are too big, let's look at a model of the stars within 13 light-years. As mentioned earlier, there are some forty stars in about twenty-nine systems within that distance. Let's create a bubble around the Solar System with a radius of 4 pc, or 13 light-years, and make the bubble just big enough to fit inside a football field, which is 300 feet in length (90 meters)—4 pc would be half a football field to the edge of the bubble, or 150 feet (45 meters). At this scale, 150 feet divided by 13 light-years equals 11.5 feet per light-year (or 3.5 meters per light-year), so with the pinhead Sun in the middle, Alpha Centauri would be 11.5 x 4.3 light-years (which equals 50 feet or 15 meters) away.

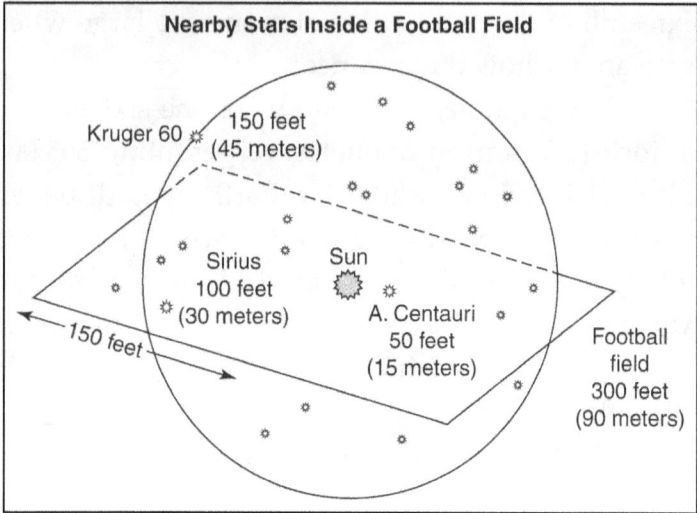

Nearby Stars Inside a Football Field

Kruger 60
150 feet
(45 meters)

Sirius
100 feet
(30 meters)

Sun

A. Centauri
50 feet
(15 meters)

150 feet

Football
field
300 feet
(90 meters)

Diagram 17: Nearby Stars Inside a Football Field

The farthest stars in our sample, Kruger 60 A and B, are (at $13 \times 11.5 = 150$ feet) at one of the goal lines. The apparently brightest star in the sky, Sirius, would be a much brighter and slightly larger pinhead at 8.6 light-years times 11.5—about 100 feet (30 meters) away.

So what we have are forty or so pinheads randomly scattered in every direction in a sphere 300 feet in diameter. All the rest is darkness.

Interestingly enough, if we expand our sample five times, to the distance of 20 pc or 65 light-years, we only come up with some 2,000 stars in a volume.

4/3 π r³ = 4/3 × 3.14 × 65³ = 1,000,000 cubic light-years, or

4/3 π r³ = 4/3 × 3.14 × 20³ = 33,000 cubic parsecs

So, if there are forty stars in 10,000 cubic light-years, there should be 1 million cubic light-years divided by 10,000 cubic light-years, which equals 116 times as many stars, or 40 × 116 = 4,600 stars.

But the Gliese catalog, a catalog of nearby stars, only lists 2,127, so half of the stars are missing—no doubt mostly red dwarfs. Even at this distance (up to 65 light-years, or 750 feet, or 250 meters or two and a half football fields), we will pick up more Sirius-like stars—Altair at 16 light-years, Vega at 25 light-years, and Folmalhaut at 23 light-years—but only 4 red giants—Pollux at 34 light-years, Arcturus at 36, the twin giants Capella at 42, and, just outside our sphere, Aldebaran, at 68 light-years away.

The more stars included in the sample, the better the calibration of their absolute magnitudes. If two of the three parameters are known—absolute magnitude, apparent magnitude, or distance—the other can readily be calculated. For example, if you know the distance (3.5 pc or 11.36 light-years) and apparent magnitude of 61 Cygni, the absolute magnitude is:

$$M = m + 5 - 5 \log d_{pc}$$
$$= 5.2 + 5 - 5 \log 3.5$$
$$= 10.2 - (5 \times .542) = 10.2 - 2.7$$
$$= +7.5$$

Above, m is the apparent magnitude and d_{pc} is the distance in parsecs derived from the parallax of .285 seconds of arc.

Apparent and absolute magnitudes, along with distances, are usually known for the nearby stars, but what about stars that are much farther away? Using an instrument attached to a telescope called a spectrograph, astronomers can match spectra of stars with known distances with dimmer stars farther away. Say you match the spectra of 61 Cygni A, which is a K5V, with a much dimmer star whose apparent magnitude is +12. Then, rearranging the formula:

$$m - M = -5 + 5 \log d_{pc}$$

or simply:

$$d_{pc} = 10^{1+.2(m-M)}$$

So, for 61 Cygni, the distance would be:

$$10^{1+.2(5.2-7.5)} = 10^{.54} = 3.4 \text{ pc} = 11.3 \text{ ly}$$

And for a twelfth magnitude star that is like 61 Cygni:

$$10^{1+.2(m-M)} = 10^{1+.2(12-7.5)} = 10^{1.9} = 79 \text{ pc} = 259 \text{ ly}$$

All this math is basically reflecting three basic principles:

1. Light falls off in the square of the distance like gravitational attraction does, so moving something three times farther away means it will be three squared—or nine times—dimmer, and when moved five times farther away, it will be five squared—or twenty-five times—dimmer.
2. The magnitude scale is logarithmic.
3. It is based on the idea that the fifth root of 100 is 2.512.

Parallax only works out to about 100 parsecs. Trying to measure a parallax shift smaller than .01 seconds of arc is almost impossible from Earth. The Hipparchus satellite can theoretically measure one milliarcsecond (.001 of a second of arc), which would mean 1,000 pc—one divided by .001—but of course the smaller the shift, the less the accuracy. This is why very rare, very distant stars like Deneb (Alpha Cygni) have distance numbers all over the map. I've seen everything from 1,500 light-years to

3,200 light-years for it. The Hipparchus catalog says it is .00229 seconds of arc, which is 1 divided by .00229 equals 436 pc, or 1,423 light-years. But if you use the distance modulus, you come up with a much different figure:

$$10^{1+.2(m-M)} = 10^{1+.2(1.25-(-8.36))} = 843 \ pc, \ or \ 2,749 \ ly$$

What this is saying is that we don't really know the absolute magnitude (which is often published as -8.36) or parallax of Deneb all that well.

We only know the distances to at best 5 percent—and usually a lot worse—accuracy, so whenever someone points out a star and says it is 653.72 light-years away, bear in mind that we don't know distances to anywhere near that level of accuracy.

To add to the complications, Deneb varies in its light output even in apparent magnitude from +1.21 to +1.29, so we can't compute its absolute magnitude or parallax accurately, and it's hard to derive even its apparent magnitude exactly. In any case, Deneb is far beyond our nearby stars model; at 11.5 feet per light-year, it would be at least 3 miles (4.8 km) away.

CHAPTER 5

THE MILKY WAY

The next step is truly an enormous one—trying to comprehend the galaxy in which the Solar System resides, the Milky Way. We have to try to envision not parsecs but kiloparsecs, (thousands of parsecs), or thousands and even tens of thousands of light-years.

We can't see the whole galaxy because we're in it, much like if you live in a city, you can only see a few blocks around where you live and maybe a few tall buildings in the distance. Because of this fact, it wasn't until the twentieth century that we discovered where we are in the galaxy—and in fact that there are lots of other galaxies. And it was only within the last ten years that the actual shape of the Milky Way was discovered.

So, what are we looking at when we look up at the band of light we called the Milky Way? Well, stars mainly—some 200 billion of them, although the human eye can only resolve about 5,000 of them at best. Big stars, little stars, hot stars and cool stars, binary stars, variable stars, stars being born and dying in clusters—and we know that at least some 1,800 of the close ones have planets. All of the extrasolar planets that have been discovered to this point are relatively close to us, so there must be countless trillions of them just in our own galaxy. (Even with his primitive telescope, Galileo was somewhat surprised to discover that most of the "milky" part of the Milky Way turned out to be stars.)

Once we have moved several hundred or thousand light-years from the Solar System, we notice that there are lots of stars grouped together in what are called open clusters. One of the nearest can be seen in the fall in the Northern Hemisphere and is in the constellation Taurus—the Pleiades or Seven Sisters (M45 in Messier's catalog).

Photo 7: The Pleiades (*STScI/Hubble Space Telescope*)

If you look at that star cluster through a pair of binoculars, it's immediately obvious that there are a lot more than seven stars there—maybe fifty or so—and large telescopes have identified more than one thousand members lying some four hundred light-years from us. (Most, but not all, of the objects seen with a telescope are stars. No matter how much telescopes were improved, some objects would not be resolved into stars.) The Pleiades is a relatively young cluster, some 70 million years old, so dinosaurs were the creatures on Earth around to witness its birth. In our nearby stars model, at 11.5 feet per light-year, M45 would be almost a mile (1.6 km) away.

Notice the three little stars that make up a triangle between the two brightest stars in the bottom of the Seven Sisters; the dimmest one is a star just like the Sun. That's how bright the Sun would be at 400 light-years (m = 10.2).

Moving out farther, we come across the closest of the rather loose star groups called associations. Unlike the open clusters, they are not at all obvious. They are much more scattered and larger. A good example is the Perseus OB3 Association.

Photo 8: The Perseus OB3 Association
(*photo by Hans Vehrenberg*)

If you look just with your eyes at Alpha Persei, the brightest star in the constellation, you only see one or two stars, but look with a pair of binoculars, and lots of them appear. Unlike the much tighter open clusters, the stars in this cluster are not as gravitationally bound, and as time goes by, they will slowly disperse. The Perseus OB3

Association is about 600 light-years away, 1.33 miles (2 km) in the model.

The third class of clusters is completely different. Clusters in this class are huge—fifty to several hundred light-years across—and contain hundreds of thousands of stars. These are the globular clusters. While there are thousands of open clusters in the Milky Way, so far less than two hundred globular ones have been identified. The closest one to us is some 7,000 light-years away in the constellation Scorpius—M4. Since it is 7,000 light-years, or 2 kiloparsecs, distant in our model, it would be 11.5 feet times 7,000, which equals 80,500 feet, or 15 miles/24 km) away.

The other big difference is that, while open clusters are relatively young and filled with O and B stars, the globulars are as old as the galaxy itself—around 12 billion years old—and populated with the very oldest stars.

What about all the nebulous objects mentioned before that are not stars? They consist of the matter that formed the stars—if an object is lit up by UV radiation from nearby hot stars, it is called an H II region; if not, it is a dark nebula, like the Coal Sack in the Southern Cross.

Photo 9: M4 in Scorpius (*STScI/Hubble Space Telescope*)

Plus there is a lot of gas, dust, and grains of silicon that cause the areas around some stars to appear bluish from reflection. In addition there are remnants of stars that have aged and died, blowing off their outer layers, usually into disk-like shapes, called (inappropriately enough) planetary nebula, and there are even some supernova remnants, like the Veil Nebula in Cygnus or the Crab Nebula in Taurus.

Practically everything in astronomy is misnamed, because people name things before they know what they're talking about—planetary nebula have nothing to do with the planets, nova are not new stars, and asteroids should be called planetoids, because they are not little stars but rather small planets.

Of course, most of these clusters in nebula are a combination of everything we have discussed. For example, the Great Nebula in Orion, visible in binoculars as M42, has an open star cluster at its core, the Trapezium, and it is surrounded by gas and dust, new stars in the process of formation, and dark absorbing matter, like the famous Horsehead Nebula. M42 lies about 1,500 light-years from the Solar System, so it is the closest star-forming region—only three miles (five km) in our model.

But remember that the light our telescopes are receiving today left there at the time of the end of the Roman Empire. In miles, that's 669,000,000 mph times 24 hours a day, which equals 1.6×10^{10} hours × 365.25 days in a year, which equals 5.86×10^{12} light miles in a year × 1,500 years, which equals 8.8×10^{15} light miles in 1,500 years, which is 8,800,000,000,000,000, some nine millions of billions of miles (14,000,000,000,000,000 km). As already noted, this is why we don't use miles when discussing the depths of interstellar space and, in a like manner, we don't use millimeters to measure distances between cities. It's time for a new scale of things.

In order to make a scale model of the Milky Way, we will have to increase the model by about fifteen times. Here's one way to envision this. Go get a one-sheet road map of the United States. The United States is about 3,000 miles across, Boston to Los Angeles. Lots of people have driven it. If you could drive nonstop—no stopping for gas, food, or rest—and make a steady 70 mph, it would take forty-two hours to do it. The Milky Way is about 100,000 light-years across, so 3,000 miles divided by 100,000 light-years equals .03 miles per light-year times 5,280 feet per mile, which equals 158 feet per light-year. To get a sense of scale, here is a list of the distances: 1 light-year equals 158 feet, a hundred light-years equals 3 miles, 1,000 light-years equals thirty miles, 10,000 light-years would equal 300 miles, and 25,000 light-years would equal 750 miles.

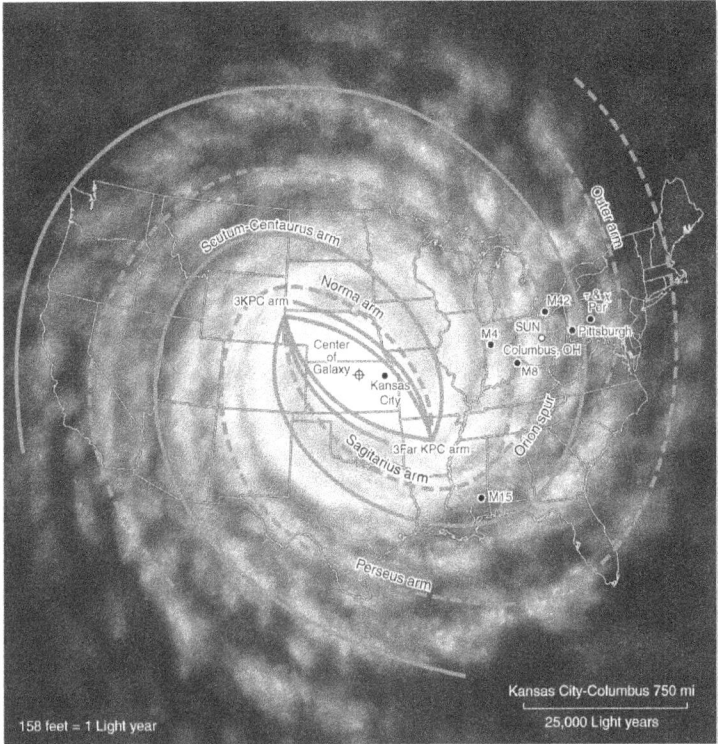

Photo 10: The Milky Way superimposed on a map of the
United States (*graphic by Steve McKinley*)

The latest map of the Milky Way shows it as
basically a two-armed, barred spiral, with
supplemental arms between them. If you
superimpose the Milky Way onto a map of the
United States, the center of the galaxy would be
where the center of the United States is, just west of
Kansas City. The Solar System would be 750 miles—
25,000 light-years—east of this spot, in Columbus,

Ohio. Almost all the stars we see without a telescope in the sky lie within about a thirty-mile radius, or 1,000 light-years, of Columbus. Toward the center of the galaxy, right about where Indianapolis is, would be the ramparts of the Sagittarius Arm, some 5,000 light-years away, and looking outward, away from the center of our galaxy, we see the trailing end of the Perseus arm, near Pittsburgh.

The Solar System actually lies between the major arms on the inside edge of what is called the Orion Spur. To get a real sense of where we are in relation to the whole galaxy, go outside on a summer evening and look at the "teapot" that makes up Sagittarius to the south. The "steam" coming out of the spout is an H II region called M8, some 5,000 light-years away in Indianapolis. If it's winter, look to the northeast for the constellation Cassiopeia—a sort of W-shaped figure—and count the stars from top to bottom—one, two, three, four—and keep going straight down toward Perseus. Halfway between, you will notice a little fuzzy patch—two star clusters called h and Chi Persei or the Double Cluster. You are now looking in almost the opposite direction of the center of the galaxy (the anticenter). The actual anticenter is a little farther south, in the direction of the star Elnath in Taurus. The two clusters are some 7,500 light-years away in the outer Perseus Arm of our galaxy—in Pittsburgh. The reason that the galaxy seems tilted is because the Solar System is inclined about 62 degrees to the plane of the Milky Way, so in the evening it's best to view the center of the galaxy on summer nights, and the anticenter during the winter. Winter is also a good time to see the Great Nebula in Orion, because you are looking down the Orion Spur some 1,500 light-years away, or about halfway between Cleveland and Columbus.

It's not really empty between the arms; there is just less stuff—i.e., dust, gas, and stars, not nothing. Since there are 158 feet per light-year, even at this scale, the nearest star, Alpha Centauri, is 4.3 light-years × 158 feet—which equals 679 feet—away.

And how big would the Solar System be? Well, the Milky Way is 100,000 light-years across— that's 365.25 days equals 36,525,000 light-days times 24 equals 876,600,000 light-hours across the galaxy. The Solar System is eight light-hours across, so 876,600,000 divided by 8 equals 109,575,000 Solar System diameters across the Milky Way. Taking 1/109,575,000 of 3,000 miles is .000027 miles times 5,280 feet per mile equals .1445 feet times 12 equals 1.73 inches! The Sun would be a tiny pinprick of light, and needless to say, the Earth has shrunk to an infinitesimally tiny speck 2/100ths of an inch from the Sun.

Europeans can make a similar model with their continent, but it's only 2,800 miles (4,500 km) even from Lisbon all the way to Moscow, plus you have to drive through nine countries, so the center of the Milky Way would be near Berlin, and the Solar System would be 1,200 km from Berlin near Kiev in Ukraine (though I'm sure it seemed a lot farther in 1941).

How about driving across the Milky Way? To drive it in our 42–hour Boston to Los Angeles trip would mean 5.86×10^{12} miles in a light-year times 100,000 light-years divided by 42 hours equals 5.869×10^{17} miles across the Milky Way divided by 42 hours equals 1.397×10^{16} mph divided by 669,000,000 mph equals 20 million times the speed of light!

Despite all the *Star Trek* movies, this is clearly impossible, since nothing with mass, whether cars or starships, can exceed the speed of light. Even *at* the speed of light, it would always take 100,000 years to cross the galaxy. In our model this means traveling 158 feet per year across the United States. A snail can move in a day at .03 mph × 24 = .72 miles in a day (3,802 feet), which is 1,387,730 feet in a year (263 miles). Dividing 3,000 miles by 263 means that the snail could cross the United States in about eleven years, so if our model of the Milky Way is superimposed onto the United States, even a snail trying to cross the United States in our model is going 9,000 times the speed of light! The traveling snail would take eleven years to get across the Milky Way, which is longer than its lifespan (five or six years), just as 100,000 years is somewhat longer than our lifespan.

I never like to say that something is impossible, because when you say that, you are really saying that you know absolutely everything there is to know about physics. For example, there was a famous professor at Harvard that declared back in the 1920s that mankind could never leave the Earth and go to the Moon. His reasoning was impeccable—the exhaust velocity of a rocket engine was, at most, 6,000 mph (9,600 kph); since escape velocity from the Earth is 25,000 mph (40,000 kph), we were clearly never leaving the Earth. What he never thought of—and neither did anyone else at the time—was the concept of the stage rocket.

Thank US Army Colonel—later Major General—Holger Toftoy for this innovation. After capturing leftover Nazi V-2 rockets from Germany, Toftoy—almost forgotten nowadays—had the idea in 1946 to attach a WAC Corporal sounding rocket to the top of a V-2. His concept was formalized as the Bumper Project in 1947, and the first stage rocket—the "Bumper-WAC"—was successfully fired in 1948.

So, while interstellar travel seems to be impossible, and at our stage of knowledge clearly *is* out of the question, who knows what discoveries and inventions the future holds? I'm sure if you went back to the time of Columbus in the fifteenth century (which, by the way, is also impossible) and told him that the day would come when people would regularly cross the Atlantic in six hours, he would find that hard to believe.

CHAPTER 6

THE GALAXIES

Now we must make another truly great leap—out to the realm of the galaxies. Until the mid–twentieth century, the Milky Way *was* the Universe. All the planets and stars and everything were considered to be part of our star system, with what was supposed to be infinite space surrounding it. There was some speculation by philosophers, namely Immanuel Kant in the eighteenth century, that at least some of the faint nebula might be other milky ways (he called them island universes) at great distances, but it was just an unproven idea; there was no scientific data one way or the other. The scientific evidence seemed to be contradictory.

In the nineteenth century, however, as telescopes got larger and larger, and photography was adapted to astronomy, it looked like stars could be resolved in the distant nebula—but at the same time astronomers like Adriaan van Maanen published data indicating that the spiral nebula were rotating at such a rapid rate that if they were at great distances, they would have to be rotating faster than the speed of light. The key lay in finding a way to derive their distances. That would not be parallax, since the nebula were much too far away—

remember parallax only works out to about 100 pc—but by the discovery of a certain type of variable star. There are about fifty different kinds of variable stars—everything from pulsating red giants to small red dwarfs that flare up, and even stars that don't really vary, but only eclipse each other as they move across our line of sight. The ones that interest us are the Cepheid variables. These are stars that are very luminous giants (so they can be seen at great distances). They pulsate not only at a regular rate but also at a rate that is based upon their intrinsic brightness or absolute magnitude. Sort of like the fact that elephants' hearts beat more slowly than do those of mice. Astronomer Henrietta Leavitt discovered this interesting and vital fact while she was looking for variable stars in the Small Magellanic Cloud.

Photo 11: The Small Magellanic Cloud
(*photo by Hans Vehrenberg*)

Now, nobody knew how far away the Small Magellanic Cloud was, but it was safe to assume that all the stars in it were at about the same distance—similar to how if you live in Los Angeles you might say that all the lights in Chicago are the same distance from you. True, the ones on the west side of Chicago are closer to you than the ones on the east side, but at 2,000 miles, the difference between the east and west sides of town (about five miles) is insignificant. Looking at the light curves in a graph showing the apparent magnitude over time of Cepheid variable stars, which are very distinctive because of their unique rapid rise to brightness and slow decline, Leavitt, working with Harlow Shapley at Harvard University, created a graph that showed that the longer a Cepheid took to go through its cycle, the brighter it was.

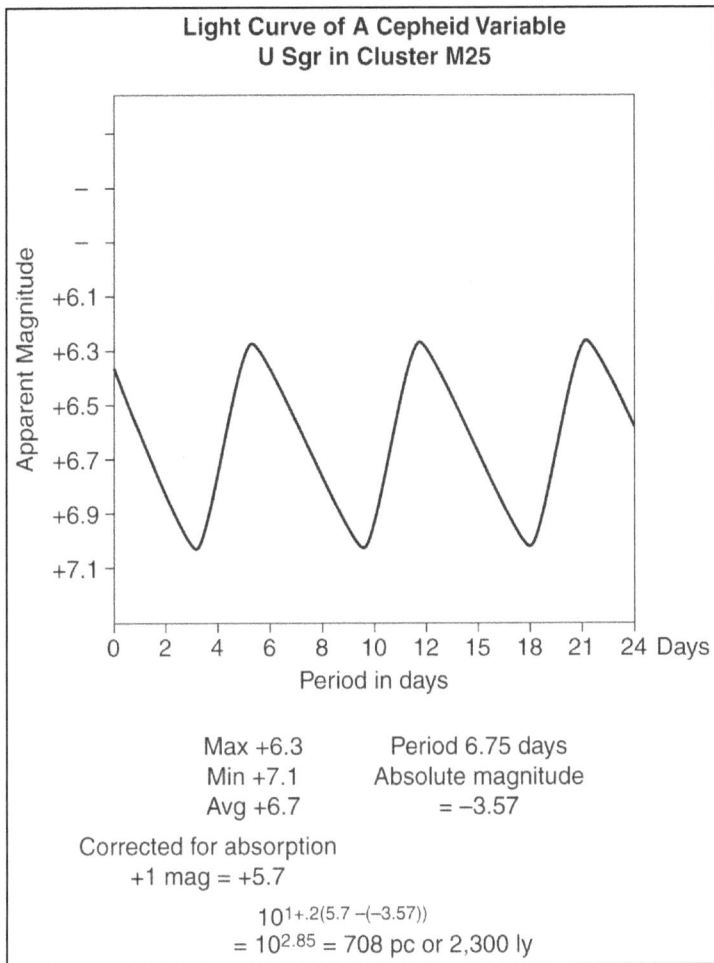

Light Curve of A Cepheid Variable
U Sgr in Cluster M25

Max +6.3   Period 6.75 days
Min +7.1   Absolute magnitude
Avg +6.7   = −3.57

Corrected for absorption
+1 mag = +5.7

$10^{1+.2(5.7 -(-3.57))}$
$= 10^{2.85} = 708$ pc or 2,300 ly

Diagram 18: Light Curve of a Cepheid Variable

This is called the period-luminosity relationship. Figuring out the periods was relatively simple—just keep taking photographic plates over and over until the star is back to its original brightness. Figuring out just how bright it was intrinsically (its absolute magnitude) was much more difficult. If only there were some Cepheid variables close enough to us to do a parallax on them and therefore have some basis in which to calibrate their absolute magnitudes—that would be great. Unfortunately, there weren't any. The closest Cepheid (not counting Polaris, which is unique in some ways) is Delta Cephei, and it's about 900 light-years from us. Actually, these days, using the Hubble Space Telescope, a parallax of .0033 has been obtained that gives a distance of 1 divided by .0033 equals 303 pc, or 988 light-years. Knowing the distance, 303 pc, and the average apparent magnitude, +4.07, it is easy to find its absolute magnitude:

$$M = 4.04 + 5 - 5log(d_{pc})$$
$$M = 4.04 + 5 - (5log\ 303\ pc)$$
$$M = 4.04 + 5 - (5 \times 2.48)$$
$$M = 9.04 - 12.4$$
$$M = -3.36$$

A very bright star indeed! Some eight magnitudes brighter than the Sun—in fact, 2.512^8 = 1,600 times as bright as the Sun intrinsically!

But back in at the beginning of the twentieth century, with no Hipparchus satellite or Hubble

Space Telescope available, this was beyond ground-based astronomy (actually astrometry). So how could it be calibrated? Well, here's one way. Say you have a bunch of 60-watt lightbulbs scattered around your backyard at night, and they are close enough to measure their distances with a tape measure. Now, you look away to a distant neighbor's yard and notice that there is a very bright pulsating ball in his yard that is clustered with some 60-watt lightbulbs. Using the inverse square law of light and assuming that the bright pulsating ball is the same distance away as the 60-watt ones in the neighbor's yard, you could figure out how bright the brighter bulb is even though there aren't any of them in your yard to measure directly. Using Cepheid variables that are in open clusters, like U Sgr in M25, a few Cepheids were calibrated. Today we know the Small Magellanic Cloud is a subgalaxy of the Milky Way laying some 200,000 light-years away.

Looking for novae—stars that suddenly brightened and then seemingly disappeared—in the Great Andromeda Galaxy M31, Hubble finally discovered a Cepheid variable, measured its period, and figured out that it was at least some hundreds of thousands of light-years away.

Photo 12: The Great Andromeda Galaxy — M31
(STScI/Hubble Space Telescope)

If it was that far away, it was another Milky Way, not a small rotating mass of gas relatively close to us that could be a solar system in the process of forming, as some astronomers thought. Distance is everything. Of course it's not as simple as I made it sound. Hubble only figured that M31 was 1 million light-years away, and now we know it is some 2.4 million light-years distant.

One of the complications was the discovery of interstellar dust and gas that dims stars over large distances. In fact, in the plane of the Milky Way, this extinction amounts to about one magnitude for every kiloparsec or 3,000 light-years. So a correction had to be added. Also, after World War II, Walter Baade, working with the great 200-inch telescope on Mount Palomar, discovered that there were two different populations of Cepheids with different absolute magnitudes.

In order to make a model of even the nearby galaxies, the so-called Local Group, it will be necessary to shrink the Milky Way down a lot. A penny or a five-cent euro, which is three-fourths of an inch in size (22 mm), will now represent 100,000 light-years. I leave it to your imagination to try and conceive just how tiny the individual 200 billion stars are that make up the galaxy at this scale. Along with the Milky Way and M31 in Andromeda, the only other fairly large galaxy in the Local Group is M33, at about the same distance as M31. The other components of the Local Group, which is usually defined as a sphere with a radius of 3 million light-years, are all dwarf galaxies—with only millions or hundreds of millions of stars, not hundreds of billions. They include both the Magellanic Clouds and such esoteric components as Leo II, the Fornax Dwarf, and the Wolf-Lundmark-Mellott Galaxy, another dwarf—some fifty galaxies in all.

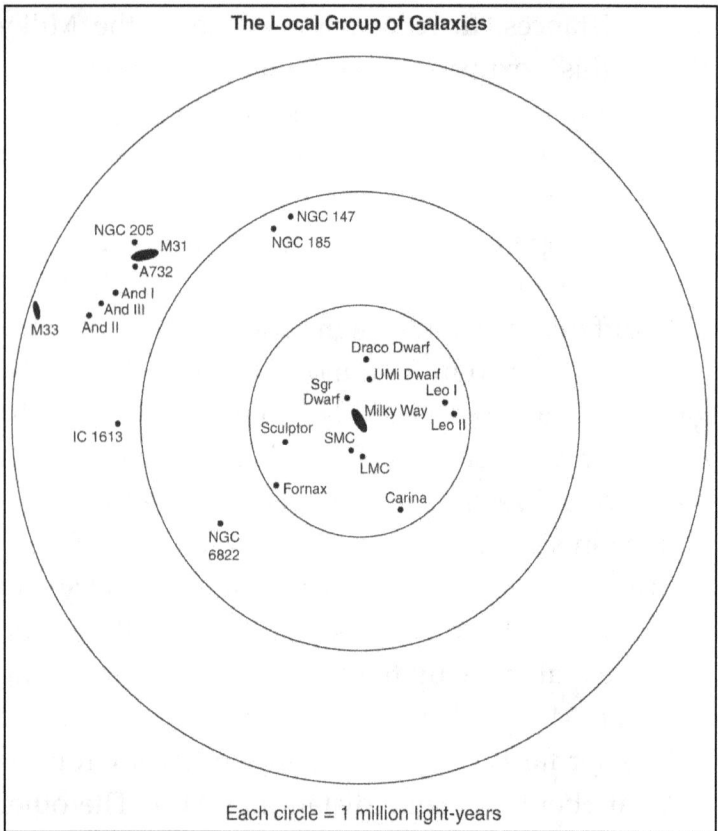

The Local Group of Galaxies

NGC 205
NGC 147
M31
NGC 185
A732
And I
And III
M33   And II
Draco Dwarf
UMi Dwarf
Sgr
Dwarf
Leo I
Milky Way
Leo II
Sculptor
IC 1613
SMC
LMC
Fornax
Carina
NGC
6822

Each circle = 1 million light-years

Diagram 19: The Local Group of Galaxies

To make a penny model where 100,000 light-years equals three-quarters of an inch, the Large Magellanic Cloud would be only 1.6 inches away and much smaller than a penny, at .14 of an inch in size (a map tack), the Small Magellanic Cloud—about that half the size of the Large Magellanic Cloud—a little farther away at 2 inches. The closest big galaxy, M31, would be 2.4 million light-years divided by 100,000 light-years equals 24 Milky Way diameters away, or .75 × 24 = 18 inches (45 cm). Since it covers about 3.4 degrees of the sky, believe it or not only the Milky Way—which of course covers 360 degrees of the sky because we're in it—and the two Magellanic Clouds are larger in apparent size. The Sun and the Moon are both half a degree. It is now easy to figure out just how big M31 is using the small angle equation: 3.4 degrees × 60 inches/degree × 60 inches/minute = 12,240 seconds of arc for the size of M31.

$$d = \frac{aD}{206,265} = \frac{12,240 \times 2,400,000}{206,265} = 140,000 \ ly$$

So—a little bigger than our Milky Way.

Using the size of the United States as a model, this is just too big. Remember, the continental United States is 3,000 miles across, which represented 100,000 light-years. The

Andromeda galaxy is 2.4 million light-years away, or 2,400,000 divided by 100,000 equals 24 Milky Way diameters. In our USA map model, this would be 3,000 miles times 24 equals 72,000 miles (115,000 km) away, over one-third of the real distance to the Moon! Galaxies within their groups or clusters are relatively close to each other in comparison to their stars. This is why you often see pictures of colliding galaxies.

Photo 13: The Colliding Spiral Galaxies of Arp 274
(STScI/Hubble Space Telescope)

Interestingly enough, stars are much farther apart compared to their sizes. For example, say the

Sun, an average-size star, is 865,000 miles in diameter. Alpha Centauri, about the same size, is 2.6 × 10¹³ miles away: 2.6 × 10¹³ divided by 865,000 miles equals around 30 million star diameters in distance from the Sun. So, even in colliding galaxies (which happen all the time—M31 is going to run into the Milky Way in about 3 billion years), it would be very hard for two stars to collide—the difference between one chance in 24 and one in 30 million. Of course, if the stars were closer together, like they are inside globular clusters, they have a much greater chance of a collision, and, in fact, this may explain the phenomena known as the blue stragglers.

Okay, now we have shrunk the Milky Way down to the size of a penny—a penny in which three-quarters of an inch in size represents 100,000 light-years. Of course, galaxies are like cities. You don't just draw a boundary, and then after the last house or building in the city limits there is just wilderness; cities just sort of peter out. So do galaxies—even more so since there are no legal boundaries, as far as we know, and so the density of the number of stars just sort of drops off.

Now in our model, every million light-years is 1 million light-years divided by 100,000 light-years or 10 Milky Way diameters, or .75 × 10 = 7.5 inches (in metric, 19 mm × 10 = 190 mm). Ten million light-years are 10 × 7.5 = 75 inches, 2 yardsticks, or 3,930 mm). In our Local Group model, a six-million-light-year bubble, it would be a sphere that is 6 million divided by 100,000 equals 60 Milky Way diameters in size, which translates into 60 × .75, or 45 inches (1,143 mm), a beach ball populated by a couple of pennies and some forty other ibuprofen-sized dwarf galaxies.

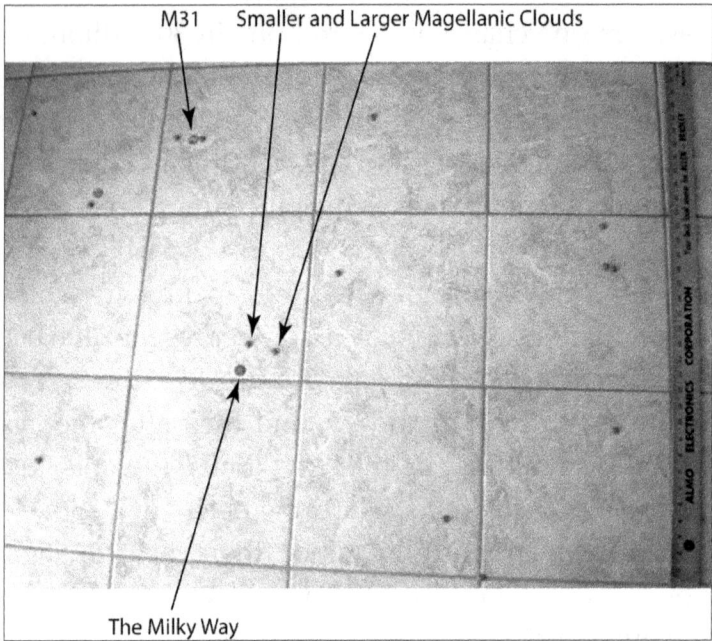

Photo 14: Coins and ibuprofen tablets representing the Local Group (*photo by author*)

In Photo 14:
Penny = 3/4″ (19 mm) = 100,000 ly
Each floor tile = 12″ (300 mm) = 1,600,000 ly

CHAPTER 7

THE REST OF THE UNIVERSE

I n order to take the next step out to 100 million light-years, we have to imagine a distance of 100 million divided by 100,000 equals 1,000 times larger than the Milky Way, so the model is now a sphere of 1,000 × .75 inches, or 750 inches (19 meters), in radius, or 63 feet—a 125-foot diameter sphere that would fit comfortably on the quarter-acre lot that my house and yard sits on. Populating this sphere are thousands of penny- and ibuprofen-sized galaxies in a volume:

volume formula 4/3 πr³ = 1.33 × 3.14 × 63³ = 1,000,000 cubic feet

About a million cubic feet, which represents a big bubble covering the property.

In intergalactic space this would equal 4/3 πr³, where 50 million light-years is the radius of the sphere, so:

1.33 × 3.14 × 50,000,000³ = 5.2 × 10²³ cubic light-years

There are some 200 small groups of galaxies within this sphere like the Local Group, but the only supercluster is the Virgo Cluster, comprising some 2,000 galaxies about 50 million light-years away. The Local Group is a small subsection of this galaxy cluster. In our backyard model, this would be 7.5 inches times 50 equals 375 inches or 32 feet (10 meters), away, just at the edge of the yard.

Beyond this are many other superclusters, some of which have many more galaxies in them than the Virgo Cluster. At the largest scale, these superclusters are embedded in a matrix of strings and voids. The largest structure in the known Universe is the Sloan Great Wall, a string of galaxies (some 11,000 of them) a billion light-years long and a billion light-years away (some 640 feet, or 200 meters, away in our model).

Photo 15: The Virgo Cluster (*STScI/Hubble Space Telescope***)**

The other really great discovery of Edwin Hubble was the fact that, as he looked at dimmer and dimmer galaxies at presumably farther distances, the spectra that Milton L. Humason took for him at the 100-inch telescope showed that they were receding from the Milky Way at faster and faster speeds.

The radial velocity, the speed that something is approaching or receding from us, of the Virgo Cluster is about 1,139 kps, about 2.5 million miles an hour. This sounds pretty fast, but it is only .4 percent of the speed of light. The more distant galaxies are moving away at increasingly faster speeds. For example, the Hercules Cluster, at 480 million light-years away (3,540 feet in our model, over half a mile), is receding at 10,300 kps (23 million mph), which is still only 3 percent of the speed of light. The graph of this relationship is simple: For every megaparsec (1 million parsecs, or 3.2 million light-years), galaxies are moving away from us at 71 kps.

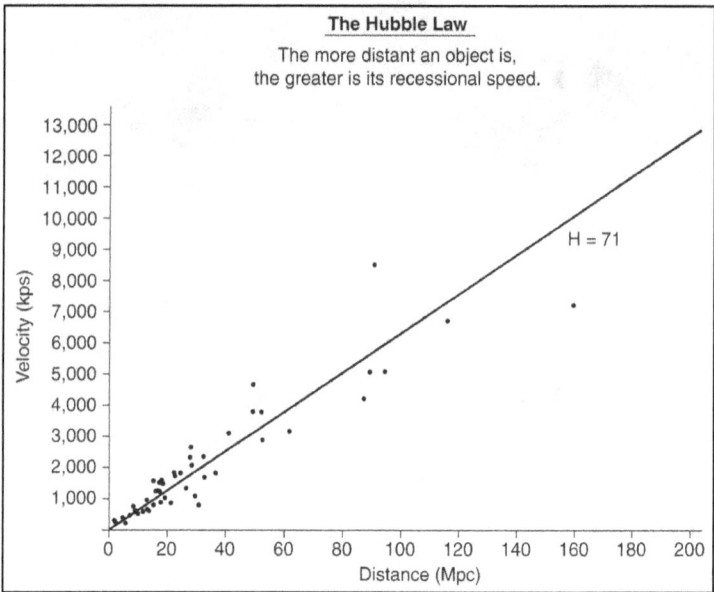

The Hubble Law

The more distant an object is,
the greater is its recessional speed.

H = 71

Velocity (kps)

Distance (Mpc)

Diagram 20: The Hubble Law

This does not mean that we are at the center of the Universe, and that everything is receding from us at 71 kps per megaparsec. In fact, this indicates that the galaxies are not really moving, except when they are close enough to interact gravitationally, but that space itself is expanding. Therefore, every galaxy would see the same phenomenon—an expanding Universe in every direction.

Does this mean that the Universe is infinite, and that the farther you look back in space, the faster galaxies are receding?

No, because at great distances (millions or billions of light-years), we are seeing things not where they are today, but where they were in the past—because the speed of light is not infinite.

One of the closest quasars (young, superluminous galaxies), 3C 273, is about 2 billion light-years away. In our model, this would be 20,000 Milky Way diameters, or 20,000 × .75″ = 15,000″, or 1,250 feet (416 meters)—about four football fields. It is receding from us at 15 percent the speed of light, or 45,000 kps, so this speed divided by 71 kps per megaparsec equals 633 megaparsecs, or 2 billion light-years.

When we look at M31 in Andromeda, we are seeing it as it was 2.5 million years ago, when Australopithecines walked the Earth.

The light we see from the Hercules Cluster left when the highest form of life on Earth (actually still in the sea) was the trilobite during Ordovician times—470 million years ago.

But if there were someone at the distance of 3C 273 who could see the Earth, what they would see would be the first multicellular life in the sea—jellyfish and algae.

This look-back time can't go on forever. When astronomers look at the most remote galaxies at the end of the cosmic dark ages—some 12 or 13 billion years ago—the speeds are approaching the speed of light, which means that the spectra are redshifted by a significant percentage of light, which is called the Z-factor. Measuring the Doppler shift for stars that are nearby is simple:

$$V = (\Delta v - rest\ v)/rest\ v$$

(V is velocity, Δv, or *delta v*, is the observed wavelength of the star, and the rest wavelength—on Earth, in a lab—is *rest v*.)

For example, a Calcium K-line at rest in the laboratory is observed at 3933.68 Å (an angstrom is one ten-billionth of a meter), and a star's spectra shows it to be at 3934.07 Å. So (3934.07 – 3933.68)/3933.68 = .000099 of c (the speed of light), and .000099 × 300,000 = +29 kps. (The plus sign means that it is receding from us.)

This is a real star—Arcturus, one of the closest red giants.

But at very high percentages of the speed of light, this formula does not work. For example, say you have a rest wavelength of 3,000 Å and an observed wavelength of 9,000 Å. Then:

$$V = (9,000 - 3,000)/3,000 = 2c \times 300,000 = 600,000 \text{ kps}$$

That means this star is receding at Z = 2c, or twice the speed of light! This is impossible. Nothing can exceed the speed of light except the Universe itself—so we must use the relativistic Doppler effect:

$$V = \frac{2z + z^2}{2 + 2z + z^2}$$

So: $2 \times 2 + 2^2/2 + 2 \times 2 + 2^2 = 8/10 = .8$ of c, or 240,000 kps, which *is* possible. It is an asymptotic relationship, since the divisor is always larger than the dividend—larger numbers will get closer and closer to c (the speed of light), but will never equal or exceed it.

Of all the tools astronomers have utilized to come up with distances—parallax, main-sequence fitting, Cepheid variables, the Tully-Fisher relation, and others—probably the most useful at great distances have been the supernovae. There are actually several different types (type *Ia*, *Ib* and *II*), but in order to simplify, let's just consider the type *Ia*. This type is the end state of massive stars in binary systems that die in a spectacular explosion. So enormous is the explosion that they are bright enough to be seen at great distances.

Remember, the absolute magnitude of the Sun is +4.8, which means that—even with the largest telescopes, which can see to apparent magnitudes down to about +27—the distance that a Sun-like star could be seen is:

$$10^{1+.2\,(m-M)} = 10^{1+.2(27-4.8)} = 10^{\,5.44}$$

That's about 275,000 pc, or about a million light-years, not counting interstellar extinction.

Type *Ia* supernovae have an absolute magnitude of –19.6. This is 4.8 – (–19.6) = 24.4—about 24 magnitudes brighter than the Sun—so 2.512^{24} equals 4 billion times brighter than the Sun, intrinsically. These supernovae should be able to be seen at distances back to the beginning of the creation of the first galaxies.

About 5,000 supernovae have been observed, which is not that many when you consider how many galaxies there are.

And just how many galaxies are there? Back in 1995, and again in 2003–2004, the Hubble telescope was aimed at a relatively empty part of the sky away from both the plane of the Milky Way and any known galaxy clusters, and exposed for a total of eleven days, in order to gather as much light as possible.

Photo 16: The Hubble Deep Field
(*STScI/Hubble Space Telescope*)

The results were truly amazing. These were two tiny pieces of the sky, one in Ursa Major and the other one in Fornax in the Southern Hemisphere. To give you an idea of just how small this area is, hold a dime up at arm's length. Roosevelt's eye just about covers an equal area 3.3 minutes by 3.3 minutes of a square degree. A square degree has 60 × 60 square minutes—3,600 square minutes—and there are 41,253 square degrees in the whole sky: 41,253 × 3,600 = 148,510,800 square minutes, and the telescope covered 3 × 3 square minutes, or about 11 square minutes, so 148,510,800 square minutes divided by 11 equals about 1/13,000,000 of the sky.

In the Hubble Ultra-Deep Field there are about 10,000 galaxies, so 10,000 × 13 million = 135 billion galaxies. And there are probably a lot more, since at these distances only the very brightest galaxies would have shown up. Remember, in our Local Group there are fifty or so galaxies but only three of any appreciable size.

Can we extend our model using a penny as the Milky Way out to the ends of the Universe? Remember, 100,000 light-years equals three-quarters of an inch. So the furthest back in time that we can see galaxies is about 13 billion years, since there weren't any before this time—or, if there were, the cosmic dark ages makes it impossible to see them.

So 13 billion divided by 100,000 equals 130,000 Milky Way diameters times .75 inches equals 97,500 inches, or 8,125 feet, equals 1.5 miles (2.4 km) for the radius of the Universe we can see. That means every galaxy, star, planet, moon, atom, and quark are all contained in a ball less than three miles in diameter. Within this three-mile-wide sphere you will have to sprinkle some 135 billion pennies ($1,350,000,000, well over $1 billion, and even more in euros) representing the galaxies. Unfortunately, most of us can't come up with that kind of pocket change.

Anyway, this three-mile sphere represents "look-back time," but the real Universe is likely to be much larger—a Universe that we will never see, since it is beyond our photon horizon. The galaxies we see at 12 billion light-years away are no longer there, since the Universe has continued to expand—and has in fact accelerated—in the last 5 billion years or so. So you might make a diagram like this:

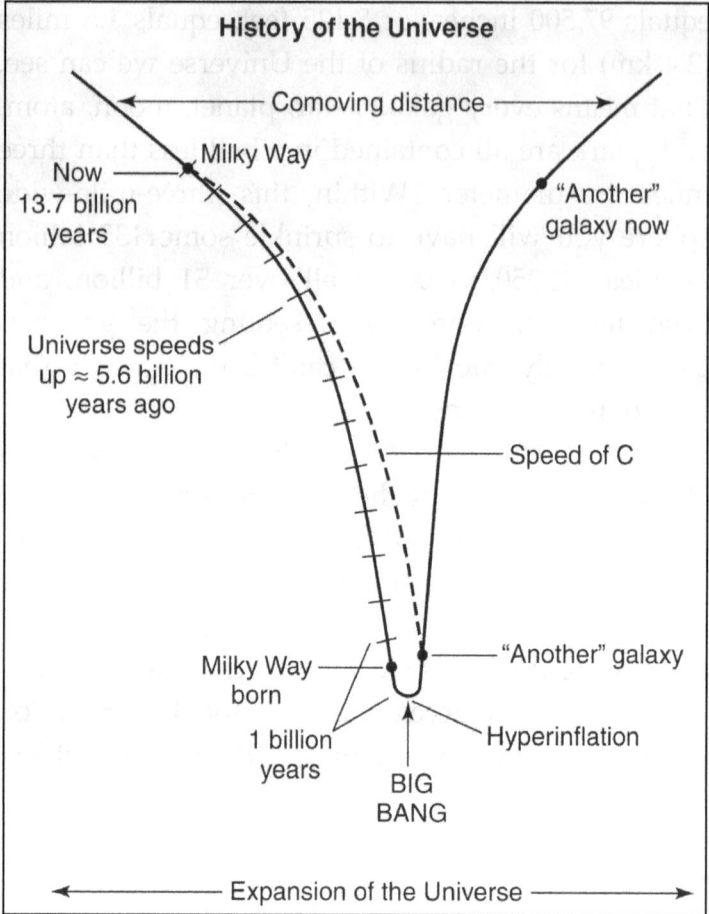

Diagram 21: History of the Universe

Cosmologists tell us that the actual size of the Universe is about 46,000,000,000 light-years in radius, so a model of the actual Universe is more like 11 miles (17.6 km) in diameter. Of course, the analogy of the big bubble can't be taken too literally, since envisioning a big bubble means you're outside of it looking at it, and there is no "outside" of the Universe—the further you look back in time and space, the less there is to look at. Go back far enough, and there was nothing to look at at all, since there *was* nothing—and in fact no events either. That's because time didn't exist, so the concept of "when" loses all meaning. Asking the question of what was going on before the Big Bang is like asking someone who is forty years old what he or she were thinking during World War I—they didn't exist so they weren't thinking anything.

The really amazing thing is that everything I've described—planets, stars, and galaxies—make up only about 4 percent of the Universe. Dark matter, which interacts only with gravity, makes up some 23 percent of the matter in the Universe, and the remaining 73 percent is theoretically tied up in dark energy, which interacts with nothing but space and seems to have made the Universe reaccelerate, starting about 5 or 6 billion years ago.

There has been much speculation lately that perhaps when our Universe was created, many other universes were made at the same time. But even if this is so, making a model of an infinite number of universes is beyond our model-making capabilities, because one cannot make a model of infinity—at least I can't.

CHAPTER 8

TIME

As challenging as it is to come to grips with the sizes and distances in the Universe, grasping the time scales involved is equally difficult to grasp.

As children we all began life completely egocentric—there was no before and there will be no after "me." A lot of people seem to retain this attitude throughout their life. My first introduction to time in history happened at a very young age. I must've been about four or five years old, and my father, who was in the US Air Force, took me to the airbase where he worked. He had to stop by a huge hangar for some reason, and he told me to wait in our 1947 Ford. Of course, after a matter of only minutes, I became bored, so I got out of the car and entered a little door into the cavernous interior. There I was, confronted with a large array of leftover warplanes from World War II, still in their wartime liveries, with nose art and bombs painted on the side indicating how many missions they had completed. B-17s, B-24s, and various fighter planes lined the hangar. I was very impressed; I didn't know what it all meant, but for the first time I became aware that "something" of great importance had happened before I was born.

I have dated my lifelong interest in history from this moment, and whether visiting a Civil War battlefield only 150 years old or the Roman Forum, some 1,800 years old, I have tried to grasp the flow of time even just on the scale of human history.

Now, World War II is hardly in the dim past, even for present-day young people. Most of us, I suspect, are able to feel how long ago the Civil War was or even a couple of hundred years back to the beginnings of the United States. The problem isn't really just the number of years, it's how much the world we are familiar with has changed. For example, we can readily imagine the number "two hundred," as in two hundred years ago, but it is much harder to imagine that the continent we live on was pretty empty. The population of the United States was 7 million, mostly confined to the eastern seaboard. It would have taken six weeks just to get from New York City to the Mississippi River, and nobody except Lewis and Clark had made the trek all the way to the Pacific. It really does sound like a different world.

Now try to go back an order of magnitude of 2,000 years, or even further—to the beginnings of recorded human history some 5,000 years ago. Most people are somewhat surprised when I tell them that we are closer in time to the Roman Republic when Julius Caesar lived than *they* were to the Old

Kingdom in ancient Egypt (2700 BC) when the pyramids were built. Even more amazing is the fact that we are closer in time to when the Tyrannosaurus Rex lived (65 million years ago) than he was to when the first dinosaurs walked the Earth (230 million years ago). The further back we go, the more unfamiliar the world becomes. Can we possibly conceive of the whole of the history of the Universe—some 13.7 billion years?

Once again we must resort to making models. One way of doing this is to make a linear model on a piece of paper—a *long* piece of paper. The old, nonadhesive shelf paper liner, 9 or 12 inches (30 cm) wide, is ideal. It will have to be about 50 feet long, since we are going to make a model where each meter (39.37 inches) equals 1 billion years, so 39.37 × 13.7 billion years equals 539 inches divided by 12 inches per foot equals 45 feet, so a 50-foot roll will do nicely. This means every centimeter will represent 10 million years (1 billion years divided by 100 centimeters per meter equals 10 million). And every millimeter will be 1 million years: 10 million divided by 10 equals 1 million years per millimeter. At this scale, it seems to us that most of the really interesting events happen during the first and last million years on our model—the first and last millimeter. Stretch out the roll of paper to the end and mark off 13.7 billion years, which is 13 m + 70 cm, then take the last centimeter and mark off the ten divisions on the millimeter scale. The last millimeter contains the Big Bang and all the basic events that happened to create the Universe we live in today. From 10^{-43} seconds after the Big Bang was the Planck era, just after the Big Bang. From 10^{-43} to 10^{-36} seconds was the grand unification epoch, when all the forces were combined. This was followed by the electroweak epoch, from 10^{-36} to 10^{-12} seconds. Then came the creation of quarks, from 10^{-12} to 10^{-6} seconds. Finally matter was created: hadrons (which are protons and neutrons), from 10^{-6} to 1 second; leptons (among which are electrons), from 1 to 10 seconds; up to the photon epoch, from 10 seconds to around 380,000 years following the Big Bang.

Here, for the first time, you can actually see enough of the scale to make a mark (if you have a really sharp pencil)—about a third of a millimeter from the beginning, which marks the three-degree background radiation that pervades the Universe in all directions. The WMAP satellite mapped this event with such precision that it actually resolved the clumpiness caused by quantum fluctuations that would become the seeds of future stars and galaxies.

All these events took place well within the first millimeter on our scale model of 13,700 mm.

What follows is called the "cosmic dark ages," from about 150 million to 800 million years after the Big Bang. There wasn't much to look at, since the Universe was foggy; this was due to the fact that the photons were interacting with the protons, neutrons, and electrons—not until the photons were decoupled could the Universe become transparent. In our model the dark ages are 150 to 800 mm forward from the end, or 15 to 80 cm.

After this there was a long period when the first stars were born (some 400 million years after the Big Bang) lived, and died, and the numerous generations of galaxies, including the Milky Way, formed into clusters and superclusters. About 9 billion years after the Big Bang—9 meters from the beginning of the roll—our Solar System was formed. The Earth and the Moon together make a convenient record of the Solar System, since about that time the Moon stopped having catastrophic events and looked more or less like it does today, whereas the Earth has pretty much erased everything from the early Solar System through the action of plate tectonics, glacial epochs, changes in sea level, mountain building, and wind and erosion—none of which happened on the Moon.

The theory is that some 4.55 billion years ago, 4 m + 55 cm back from the present-day end of the roll, when there were a lot of large bodies orbiting the Sun in very unstable orbits, a Mars-sized object hit the partially differentiated Earth and blasted out enough matter to form the Moon. The heavy stuff like the nickel and iron had already settled to the core of the Earth, so only the lighter stuff—the silicate rocks—got blasted off, which is why the Moon is one-quarter the size of the Earth but the Earth has eighty-one times more mass.

At first the Moon was very close to the Earth, but not closer than 12,000 miles—that is the Earth's Roche limit. Had the Moon been any closer than that, it would have formed a ring around the Earth, not a moon. The Moon steadily retreated back from the Earth at a rate of about 1.5 inches (38 mm) per year.

Believe it or not, the Earth and the Moon looked pretty much the same at this time—semimolten bodies slowly cooling while large asteroids continuously slammed into them, forming the big, multiringed basins on the Moon. About 3.8 meters back from the present-day end of the roll, the great Imbrium impact event took place, creating one of the larger craters in the Solar System on the Moon. After this, things pretty much calmed down on the Moon except for an occasional asteroid impact that we can see today as large craters on the Moon. The "youngest" craters are the ones that have rays.

The oldest rocks on Earth have been dated to about 3.8 billion years ago. Our planet, however, would be completely unrecognizable to us as it looked back then. Small isolated continents were emerging from the sea and contained the first microfossils, some 2.8 meters back from the end. Measure 60 cm back from the present day (600 million years)—this marks the time when the Earth starts to really get interesting (whereas the Moon looked almost the same as it does today, as I noted above). The so-called Cambrian explosion of life that suddenly fills the fossil record with all kinds of multicellular life—including trilobites, corals, sponges, and mollusks—dates from about 540 million years ago, or 54 cm back in our scale. It's interesting to consider that the Moon would have looked about as it does today to the first creatures that could see, the trilobites (if they had looked at it), except that it was a lot bigger in the sky—one degree at 115,000 miles (184,000 km) away. So the last 13.7 billion years—13 m + 70 cm, or 13,700 mm, on our sheet—look like this, with some of the major events included:

TIME SCALE OF THE UNIVERSE

Years before present day	Measurement back from present day	Events on Earth/Moon	Look-back time to objects
Today to 1 million years ago	1 mm	All of history	M8 in Sagittarius Arm and everything else in the Milky Way
2,500,000 years ago	2.5 mm	First humans	M31 in Andromeda
50,000,000 years ago	50 mm	Breakup of Gondwana Land, Tycho Crater	Virgo cluster
65-250 million years ago	65-250 mm	Reign of the dinosaurs	
250-360 million years ago	250-360 mm	Permian extinction	Coma cluster
500 million years ago	500 mm (50 cm)	Ordovician Times: Trilobites	Herculus cluster
600 million years ago	60 cm	Cambrian explosion of life	
800 million years ago	80 cm	Copernicus crater	
1 billion years ago	1 meter		Sloan Great Wall
3.8 billion years ago	3.8 meters	Imbrium event	
4.6 billion years ago	4.6 meters	Formation of Earth, Moon, & Solar System	
6 billion years ago	6 meters		Universe starts to accelerate
6-11 billion years ago	6-11 meters		Formation of Galaxy clusters
13 billion years ago	13 meters		Milky Way forms
	Last mm		All subatomic particles & laws of the Universe created

Table 2: Time Scale of the Universe

Now go to the very beginning of the scroll—the last millimeter represents 1 million years, so this tiny division encompasses not only all of human history, but just about all of humanity. (The first really manlike creature, *Homo erectus*, dates from about 1 million years ago.) So we have to increase the scale by tenfold in order to see the events of the last million years:

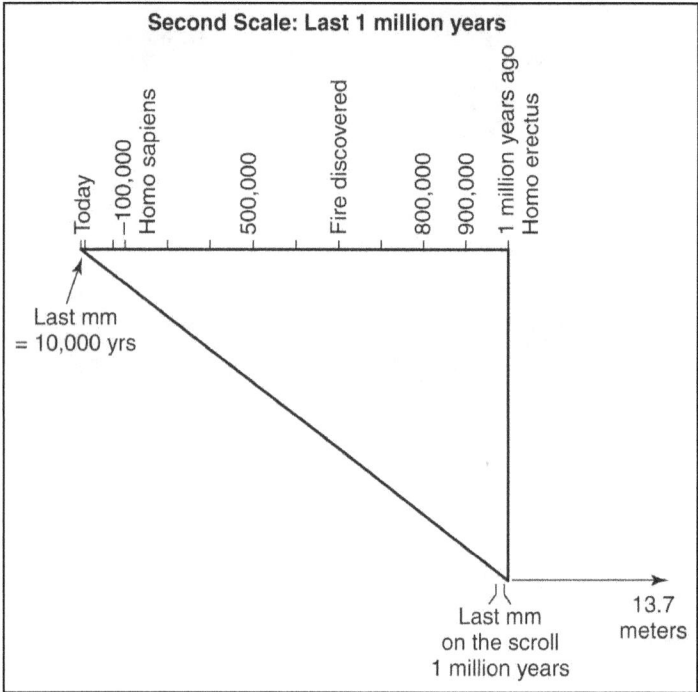

Table 3: The Last 1 Million Years

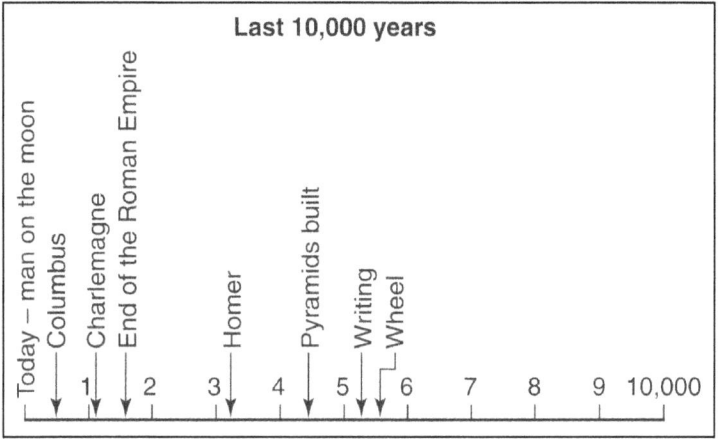

Table 4: The Last 10,000 Years

Now each centimeter (10 mm) equals 100,000 years, and each millimeter equals 10,000 years. Since the last millimeter on the scale equals 10,000 years—double all of human history—we must once more blow up the scale by ten times—now one millimeter equals 100 years. At this scale the last 10,000 years is 10 cm. (Remember, both of these two scales fit inside the last millimeter on our scroll.) Looking back you get at least some sense of the true scale.

I think it is really ironic that dinosaurs have always been pretty much portrayed as *the* iconic failures in the history of life ("Why are you still driving that dinosaur of a car from the 1970s?"). Looking at the scale of time, dinosaurs were around from Triassic through Cretaceous times—160 million years—whereas human beings from the most primitive to today only span about 2 million years. The lifetime of the average species on Earth is around 5–10 million years (mammals much less, maybe 1 or 2 million years), with some obvious exceptions—like the horseshoe crab, which has been with us virtually unchanged for the last 400 million years. So I really don't know how to answer the question I am always confronted with when I tell students that the Sun is going to die in about 4 billion years, and they inevitably say, "But what will people do when that happens?" I think it's pretty safe to bet that no type of recognizable humans will be around to witness the death of the Sun.

CHAPTER 9

GO OUT AND LOOK

Theree is an old joke in astronomy about how to tell the difference between amateur and professional astronomers: Professionals spend way more time studying their telescopes than they do the sky. Even the astronomical instrumentation used by amateurs has become so complicated that it is not a unusual for fifty people to show up in a field and spend an hour and a half setting up their telescopes and cameras.

Go-to telescopes and computerized imaging have their place; however, they will never teach you about astronomy. No Discovery Channel DVD or computer program will take the place of going out and seeing for yourself. Practically everyone has seen the fabulous Voyager and Cassini spacecraft pictures of Saturn, but it has always amazed me how, when I would show people the rings of Saturn—even in the smallest telescopes—they are always astonished. I guess that it has little to do with the image and much more to do with human psychology. (A common comment is, "It's really there!")

Astronomy is not like other hobbies—it has to be done at the sky's convenience. For example, say it's Saturday night, and you don't have anything to do, so you have the opportunity to work on your stamp collection. This just does not work for astronomy. Not only do the skies have to be clear—astronomical events are not necessarily scheduled for your convenience. If you want to see sunrise on Mount Pico on the Moon, you only get one chance per month. The transit of Venus in 2012 was the last chance to see Venus cross the Sun until 2117.

Astronomical events are usually known far in advance, however. I remember one time a TV meteorologist announced that there was a "chance of a lunar eclipse tonight." Well, that is the difference between meteorology and astronomy— the lunar eclipse *was* going to happen. I think he meant that there was a chance that it would not be visible because of cloud cover.

So what do you need in order to enjoy the sky? I would advise people to get a planisphere (or the phone app equivalent), then go out in the yard, away from outdoor lighting, and learn the motions of the celestial sphere first. (Remember, cavemen built Stonehenge with practically no technology, but with a real understanding of the apparent motions of the Sun and Moon.) The only technology needed to demonstrate the seasons is a nail and a piece of wood that is not moved for a year (and patience— something in short supply in the modern world, but seemingly abundant long ago). In this photo, the little nails were pounded in the railing at the solstices and equinoxes.

Photo 17: The Sun's Shadow near the Winter Solstice
(*photo by author*)

The next step is to go out and learn all the bright stars and constellations that you can identify. Only after you have accomplished this should you invest in binoculars or a small telescope. I remember once when I was first starting to teach astronomy, a man came up to me about halfway through the class and said that he had purchased a telescope but couldn't find anything in it. It turns out that he had a lot of money and had bought a 14-inch Celestron that was far beyond his knowledge to operate. One would not try to learn to sail a boat by taking on a sixty-foot schooner (a Sunfish would be a better choice). People often ask me what kind of telescope to buy. Well, that's like asking someone what kind of car to buy. Someone with five children will have to buy a different car than someone who is single. It's the same thing with telescopes—if you're interested in looking at the Moon and planets, a 3- or 4-inch refractor will do nicely, but if you're interested in galaxies and deep-sky objects, then you should invest in as big a reflector as you can handle, are willing to take outside—and can afford. I remember going through a period of time when I kept buying bigger and bigger telescopes. Finally I had a 12-inch Cave Newtonian reflector that was so awkward and clumsy (and heavy) that it was never *quite* clear enough to go through the trouble of hauling it out, so the telescope seldom got used. If it is sitting indoors, even the best telescope is of no use whatsoever.

The Moon

Even with a small telescope, our Moon affords a lifetime of study and entertainment. I have been observing the Moon for about thirty-five years, and I am still discovering new things all the time. Mountains, rills, domes, and uncountable craters are well within the capabilities of a small telescope, but it's difficult to get the scale of things. I know that when they first look, most people see "little holes on the Moon," and don't realize how large some of the craters are. If there was a building or truck next to one, you could get a sense of scale. Clavius, which is 232 km (145 miles) in diameter—the distance from New York City to Philadelphia, or Rome to Florence—is something that we just don't have the equivalent of on Earth. I've seen things like two-mile-diameter craters with my 4-inch refractor, but as I said, I still haven't seen it all. (It takes over two hundred years for the exact same lighting conditions to occur on the Moon because of the changing phases and librations). Every time there's a full moon, people always say, "You must be going out with your telescopes." Actually, that's the worst time to look at the Moon (unless you're interested in the rays), because the lighting conditions make it hard to see relief on the Moon when the Sun is directly overhead, like so . . .

Photo 18: Moon Shadows—Sun Overhead (*photo by author*)

. . . than it is when the Sun is shining on the Moon at an angle:

Photo 19: Moon Shadows—Sun at Angle (*photo by author*)

The Planets

The planets are all much smaller than the Moon in apparent size. Jupiter is only 1 minute in size (60″) at its opposition (closest to Earth), which is one-thirtieth the size of the Moon to the unaided eye. With a telescope at 50 power, Jupiter will appear the same size as the naked-eye Moon—with good conditions, you can expect to see some bands and of course the four Galilean moons. The more you look, the more you can see. Many times I have had twelve-year-old kids look through my telescope and not be able to see the bands—even though I know they have much better vision than I. Part of learning the sky is improving your observing skills. The more you look, the better you get at it.

At least 30 power is needed to see the rings of Saturn, and even the smallest telescope or binoculars will show the giant moon Titan—pretty incredible when you think that Saturn is almost a billion miles away.

You cannot see the surface of Venus since it's covered completely by CO_2 clouds, but you can follow the phases as it goes around the Sun, appearing first as an evening star, then a morning star.

Mercury is so tiny that it's a challenge just to find it in the dusk or dawn, since it is never more than 27 degrees from the Sun.

Uranus and Neptune, at 4 and 2 seconds in size, are so far away that, while you can identify them as disks and not stars with a very small telescope, they are not much to see.

As for Pluto, I've only seen it once or twice after hunting for about an hour, and only so I could say that I've seen *all* the planets.

Don't neglect to occasionally observe the Sun. As I'm sure you all know, never, *never* attempt to look at the Sun with just your eyes, or with a telescope, or binoculars, or anything else. You will be blinded immediately! (Remember the burning pencil?) There are safe ways to set up a telescope in order to project the Sun's image onto a screen, or you can buy an Inconel metal filter from Thousand Oaks Optical that allows only 1/400,000 of the light through. With these techniques, you can see sunspots and plages (white spots) on the surface, measure their sizes, and calculate the rotation period of the Sun. In order to see the prominences, a special hydrogen alpha filter or special telescope is needed. I never really knew many solar astronomers—I guess because they were awake during the day, when I was fast asleep.

The Milky Way

Even with my 60 mm refractor, I've seen all the Messier objects (but never in one night, although that's possible). In addition, there are the innumerable binary and variable stars to observe. And don't forget that one of the best ways to observe the Milky Way is with a lawn chair, a pair of binoculars, and just your eyes.

Galaxies

I like to show people galaxies. There's really not much detail to see using a small telescope—they appear as small blobs of light—but seeing them gives you a real appreciation of why it took until the twentieth century to discover where we were in the Universe. Also it's really amazing to think that when you're looking at them (for example, the twin galaxies M84 and M86 in the Virgo Cluster), you're seeing them as they appeared when dinosaurs walked the Earth some 60 million years ago.

When you are first learning the sky it's best to approach it as an experience. I still remember reading a Robert Burnham Jr. (author of *Burnham's Celestial Handbook*) interview that was in *Astronomy* magazine in March 1982 (when I was just learning about astronomy) where he made the point that the most important thing for everyone in astronomy is the direct experience of the Universe. How many people would go to the Grand Canyon if they were only interested in geology? True, some people may become interested in geology from visiting, but most people are mainly going there for the experience of discovery. He also made the point—and I have never forgotten this—that trying to treat the Universe like it's a file of data to be computerized and analyzed is like trying to describe a great work of art by limiting its description to a chemical analysis of the paint. The Universe is not just for astronomers and cosmologists, as much as they have taught us, but for everyone to enjoy and wonder about.

BIBLIOGRAPHY

Bartusiak, Martha. *The Day We Found the Universe*. 2009. New York: Pantheon Books.

Burnham, Robert Jr. *Burnham's Celestial Handbook: An Observer's Guide to the Universe Beyond the Solar System*. 1978. New York: Dover Publications, Inc.

Chandler, David. *The Night Sky™: The Original 2-sided Planisphere, Small 30°–40° North Latitude*. 1992. Springville, California: David Chandler Company.

Chapman, David M. F., ed. *Observer's Handbook 2013, 105th Edition*. 2012. Toronto: The Royal Astronomical Society of Canada.

Ferguson, Kitty. *Measuring The Universe: Our Historic Quest to Chart the Horizons of Space and Time*. 2000: New York: Walker Publishing Co.

Galilei, Galileo. *Sidereus Nuncius or The Sidereal Messenger*. Translation with introduction, conclusion, and notes by Albert Van Helden. 1989. Chicago: University of Chicago Press.

Gott, J. Richard and Vanderbei, Robert J. *Sizing Up the Universe: A New View of the Cosmos*. 2010. Washington, DC: National Geographical Society.

Helden, Albert Van. *Measuring the Universe: Cosmic Dimensions from Aristarchus to Halley*. 1985. Chicago: University of Chicago Press.

Hirshfeld, Alan, *Parallax: The Race to Measure the Cosmos*. 2001. New York: W. H. Freeman and Company.

BIBLIOGRAPHY

King, Henry C. *The History of the Telescope*. 1979.
  New York: Dover Publications.
Kitt, Michael T. *The Moon: An Observing Guide for
  Backyard Telescopes*. 1992. Waukesha, Wisconsin:
  Kalmbach Publishing Company.
Macdougall, J. D. *A Short History of Planet Earth:
  Mountains, Mammals, Fire, and Ice*. 1996. New York:
  John Wiley & Sons, Inc.
Norton, Arthur P., *Norton's 2000.0 Star Atlas and
  Reference Handbook, 18th Edition*, edited by Ian
  Ridpath. 1989. New York: John Wiley & Sons, Inc.
Ottewell, Guy. *The Astronomical Companion, 14th
  printing*. 1995. Greenville, South Carolina: The
  Astronomical Workshop, Furman University.
Singh, Simon. *Big Bang: The Origin of the Universe*.
  2005. New York: HarperCollins Publishers.
Tirion, Wil and Skiff, Brian. *Bright Star Atlas 2000.0*.
  2000. Richmond, Virginia: Willmann-Bell, Inc.
Vehrenberg, Hans. *Atlas of Deep Sky Splendors*. 1983.
  Cambridge, Massachusetts: Sky Publishing Corp.
Webb, Stephen. *Measuring the Universe: The
  Cosmological Distance Ladder*. 1999: Chichester, UK:
  Praxis Publishing Ltd.

Index

www.ingramcontent.com/pod-product-compliance
Lightning Source LLC
Chambersburg PA
CBHW071836200326
41519CB00016B/4136